ADOLPHE QUETELET, SOCIAL PHYSICS AND THE AVERAGE MEN OF SCIENCE, 1796–1874

SCIENCE AND CULTURE IN THE NINETEENTH CENTURY

Series Editor: Bernard Lightman

TITLES IN THIS SERIES

1 Styles of Reasoning in the British Life Sciences: Shared Assumptions, 1820–1858
James Elwick

2 Recreating Newton: Newtonian Biography and the Making of Nineteenth-Century History of Science
Rebekah Higgitt

3 The Transit of Venus Enterprise in Victorian Britain
Jessica Ratcliff

4 Science and Eccentricity: Collecting, Writing and Performing Science for Early Nineteenth-Century Audiences
Victoria Carroll

5 Typhoid in Uppingham: Analysis of a Victorian Town and School in Crisis, 1875–1877
Nigel Richardson

6 Medicine and Modernism: A Biography of Sir Henry Head
L. S. Jacyna

7 Domesticating Electricity: Technology, Uncertainty and Gender, 1880–1914
Graeme Gooday

8 James Watt, Chemist: Understanding the Origins of the Steam Age
David Philip Miller

9 Natural History Societies and Civic Culture in Victorian Scotland
Diarmid A. Finnegan

10 Communities of Science in Nineteenth-Century Ireland
Juliana Adelman

11 Regionalizing Science: Placing Knowledges in Victorian England
Simon Naylor

12 The Science of History in Victorian Britain: Making the Past Speak
Ian Hesketh

13 Communicating Physics: The Production, Circulation and Appropriation of Ganot's Textbooks in France and England, 1851–1887
Josep Simon

14 The British Arboretum: Trees, Science and Culture in the Nineteenth Century
Paul A. Elliott, Charles Watkins and Stephen Daniels

15 Vision, Science and Literature, 1870–1920: Ocular Horizons
Martin Willis

16 Popular Exhibitions, Science and Showmanship, 1840–1910
Joe Kember, John Plunkett and Jill A. Sullivan (eds)

17 Free Will and the Human Sciences in Britain, 1870–1910
Roger Smith

18 The Making of British Anthropology, 1813–1871
Efram Sera-Shriar

19 Brewing Science, Technology and Print, 1700–1880
James Sumner

20 Science and Societies in Frankfurt am Main
Ayako Sakurai

21 The Making of Modern Anthrax, 1875–1920: Uniting Local, National and Global Histories of Disease
James F. Stark

22 The Medical Trade Catalogue in Britain, 1870–1914
Claire L. Jones

23 Uncommon Contexts: Encounters between Science and Literature, 1800–1914
Ben Marsden, Hazel Hutchison and Ralph O' Connor (eds)

24 The Age of Scientific Naturalism: Tyndall and his Contemporaries
Bernard Lightman and Michael S. Reidy (eds)

25 Astronomy in India, 1784–1876
Joydeep Sen

26 Victorian Literature and the Physics of the Imponderable
Sarah C. Alexander

Forthcoming Titles

Victorian Medicine and Popular Culture
Louise Penner and Tabitha Sparks (eds)

ADOLPHE QUETELET, SOCIAL PHYSICS AND THE AVERAGE MEN OF SCIENCE, 1796–1874

BY

Kevin Donnelly

University of Pittsburgh Press

Published by the University of Pittsburgh Press, Pittsburgh, Pa., 15260
Copyright © 2016, University of Pittsburgh Press
All rights reserved
Manufactured in the United States of America
Printed on acid-free paper

Cataloging-in-Publication is available from the British Library

ISBN 13: 978-0-8229-6608-1
ISBN 10: 0-8229-6608-5

CONTENTS

Acknowledgements	ix
List of Figures	xi
Introduction: Two Average Men	1
1 Life in the War: The End of Enlightenment in Belgium, 1796–1823	19
2 Casualties of War: Quetelet and Friends in Ghent and Brussels, 1815–23	43
3 Stoking the Sacred Fire: The Administration of Observation in the United Kingdom of the Netherlands, 1822–30	65
4 From Brussels to Europe: The Creation of a Scientific Network, 1823–9	87
5 *Physique Sociale*, 1825–35	111
6 The Other Average Man: *L'Homme Moyen* and its Critics	135
Conclusion: The New Argonauts	159
Epilogue: The Average Enlightenment	167
Works Cited	169
Notes	185
Index	215

ACKNOWLEDGEMENTS

This book would have been impossible to write without the contributions of many outstanding women and men. Mark Hulliung provided the initial motivation to examine the legacy of the Enlightenment, and along with Govind Sreenivasan and Eugene Sheppard was instrumental in shaping the early form of this book. The idea to investigate Quetelet's life was inspired by a graduate seminar with Peter Buck at the Harvard Extension School, and I was particularly fortunate to be able to take a course on the history of science with Debbie Weinstein when the book was in its earliest drafts. I also thank Paul Jankowski, Alice Kelikian and the faculty, students and administrators of the Comparative History Program at Brandeis for creating the encouraging, liberating and supportive conditions under which this book first began.

At Brandeis, Ian Hopper, Shefali Misra, Claudia Schaler, Surella Seelig and Aaron Wirth provided an ideal group of friends and colleagues with whom to reflect on European history, philosophy and politics (and much else). Just as Quetelet's many projects would have been impossible without the friends and colleagues around him, so too did this book draw on the strength and joy of friendship.

Ian also deserves thanks for reading through many drafts, and for helpful debates and suggestions along the way. Timothy Wyman McCarty also commented on early drafts, and helped me to think a bit more about the difference between the book I thought I was writing and the one I was actually writing. The book also benefited from two dissertation seminars led by Eugene, Dan Kreider, ChaeRan Freeze and Karen Hansen. My thanks also go to the members of these seminars for their frank and open discussion of early drafts.

Rebecca Makas took on the unenviable task of reading through the complete manuscript multiple times, once aloud, and is responsible for many a sentence being far more clear and direct than it was on first execution. Anonymous reviewers at Pickering & Chatto, *History of Science* and the *British Journal for the History of Science* also helped to expand the scope of the secondary material and helped to make many of the themes of the book more explicit. I also thank those journals for permission to reprint material from two articles: 'On the Boredom of Science: Positional Astronomy in the Nineteenth Century', *British Journal for the*

History of Science, 47 (2014), pp. 473–503 and 'The Other Average Man: Science Workers in Quetelet's Belgium', *History of Science*, 52 (2014), pp. 401–28. Robert Smith first suggested the press during a meeting of the History of Astronomy Workshop at Notre Dame, Mark Pollard encouraged the manuscript submission, and Bernard Lightman agreed to include it in the Science and Culture in the Nineteenth Century Series as well as edited the final version of the book. I thank all of them for their work and support in bringing this book to press. In spite of all of this help, any and all errors that remain are my responsibility alone.

The majority of the research for the book was able to be completed because of the generous assistance of the Belgian American Educational Foundation, which provided for a year to study, research and write in Brussels. I thank the BAEF director Emile Boulpaep, as well as Jean-Luc de Paepe and Clair Pascaud of the Académie royale de Belgique. Closer to home, I thank Roy Goodman and the staff of the American Philosophical Society in Philadelphia, Sean Casey and the staff of the Boston Public Library Rare Books Room, Kerry Magruder and Joann Palmieri and all the folks at the University of Oklahoma History of Science Collections and the library staff at the University of Wisconsin Memorial Library. Gaston Demarée of the Koninklijk Meteorologisch Instituut van België was particularly helpful in locating images and documents on the Brussels Observatory. Research support to travel and write the book was also provided by fellowships and grants from the Andrew W. Mellon Foundation, the American Philosophical Society, the History of Science Collections at the University of Oklahoma, the Friends of the University of Wisconsin-Madison Libraries and Brandeis University. Alvernia University also provided very valuable time and resources to help bring this book to its completion.

Finally, I owe more gratitude than it is possible to express here to my parents – Mom, Dad, Joan and Nancy. Thank you for keeping me company in Brussels, and for your love and support these many years.

LIST OF FIGURES

Figure I.1: Average man (detail)	3
Figure I.2: Quetelet statue and Académie royale	18
Figure 3.1: Observatoire royal de Bruxelles	67
Figure 3.2: Observatoire royal de Bruxelles, *plan général*	74
Figure 4.1: Astronomy section of the *Correspondance*	100
Figure 4.1: *Loi de la mortalité*	118
Figure 5.2: *Qualités physiques*	128
Figure 6.1: Propensity for crime and literary ability	140
Figure 6.2: *Tableau de la croissance*	146

INTRODUCTION: TWO AVERAGE MEN

Progression to the Mean

In the last years of his life, Adolphe Quetelet (1796–1874) compiled *Sciences mathématique et physique chez les belges* (1866), a book he hoped would explain how his home country of Belgium had regained its status among the great scientific nations of Europe. Quetelet had been born into a difficult time in Belgian intellectual life, and he believed that the sciences in particular had struggled in the eighteenth century, first under the benevolent stagnation of Hapsburg oversight and then under the ruinous invasion and occupation by French revolutionary forces. While Belgian scientific life remained inert, however, he saw what he called 'an intellectual innovation of great importance, an innovation which perhaps has not been so noticed'. It was an 'innovation' Quetelet believed to be the key to Belgium's return to scientific prominence:

> The man of talent, in certain cases, ceased to act as an individual and became a fraction of the body that attained the most important results.[1]

Just a page later he reinforced the point that 'in the sciences of the new era ... *savants* have ceased to act as individuals'.[2] Quetelet was comparing the new form of science to the great 'geniuses' who he believed had driven scientific progress in Europe during the seventeenth and eighteenth centuries, a period when Belgium 'did not have the strength to take part'.[3] The book was intended as a corrective to this glum history, and it contained histories, *eloges* and recollections of the *savants* who had managed to elevate Belgian science and industry to the level of the great powers of Europe.[4]

For the reader who knows Quetelet best from his theoretical *l'homme moyen* (average man), the words above may have some resonance. The notion of an 'individual' as merely a 'fraction' of a larger body seems quite similar in fact to Quetelet's infamous Average Man, a statistical composite of individual traits that Quetelet believed would be central to a new science he called *physique sociale* (social physics). Quetelet had hoped for a quantitative science that would allow researchers to count, measure and predict human actions, and the Average Man was to be the 'centre of gravity' against which such predictions could be made. The beauty of the average can be seen (in Figure I.1) as a kind of perfect physical and moral being. Quetelet had such high hopes for this imagined man that it led to one of the most seemingly bizarre statements ever made in the history of statistics and the social sciences:

> If the average man were perfectly determined, one could ... consider it as a kind of beauty. All that deviated [*s'éloignerait*] from what it resembled ... would constitute deformity and disease. That which did not resemble it ... would be monstrous.[5]

Contemporary critics derided the idea that deviation from the average was 'monstrous', and readers both sympathetic and hostile towards social physics have tried to make sense of it ever since. One possible theory is that such excitement for averages was an extension of an Enlightenment belief in egalitarianism, and indeed much of Quetelet's social physics picked up where the social mathematics of the 'last *philosophe*' Condorcet left off. Yet seen in combination with Quetelet's reflections on Belgian men of science, the above quote suggests that the idea of composite averages, of individuals 'ceasing to exist' as they become a 'fraction of the larger body', was a theme Quetelet returned to often, whether it was in the abstract realm of social physics or the practical realities of making Belgian *savants*. As this book demonstrates, in order to make progress in social physics, both the sciences of man and the men of science needed to strive to be average.

The primary goal of this book then is to present an analysis of Quetelet's professional scientific work and thought in the creation of these two kinds of average men. Quetelet worked during an era of specialization in the sciences, yet his interests cannot easily be contained within a few disciplines. Even avoiding anachronistic labels of Quetelet – of which sociologist, criminologist and climate scientist are all possible – his work covered an enormous range of topics in the natural sciences and what would become the social sciences. It would be possible to write the life of Quetelet as the story of a statistician, an astronomer, an institution builder, a mathematician, a social theorist, an educator, an economic theorist or even a frustrated poet. All of these narratives would be possible, and indeed some have been realized, but all would be incomplete.[6]

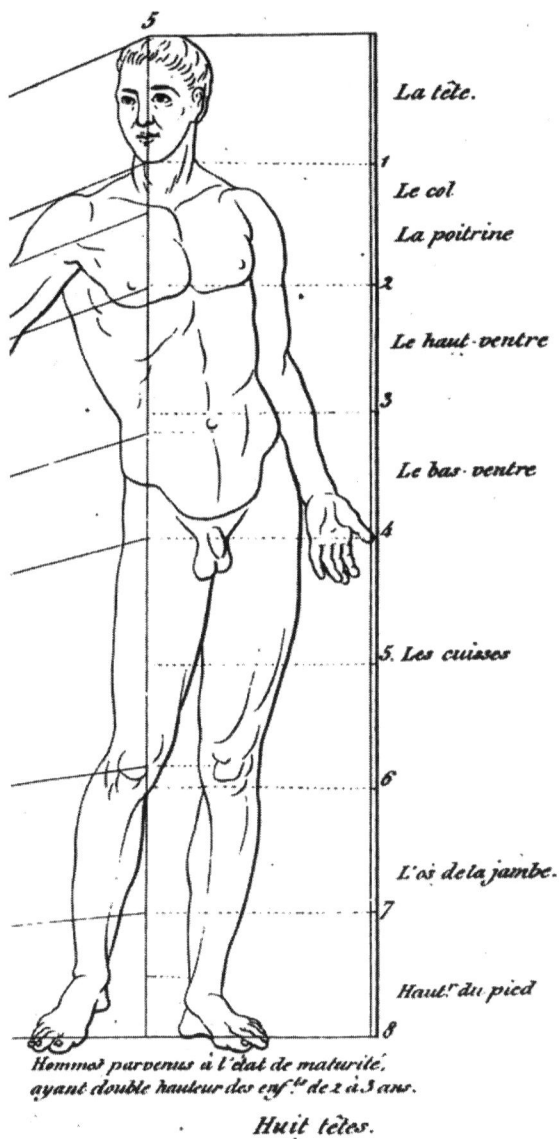

Figure I.1: Average man (detail). This image demonstrates the harmony of Quetelet's ideal average man. Although the image depicts only a physically average man, Quetelet held out equal hopes for establishing an average man of moral and intellectual character. A. Quetelet, 'Sur le poids de l'homme aux différens ages', *Nouveaux mémoires de l'Académie royale des sciences et belles-lettres de Bruxelles*, 7 (1832), pp. 1–44. Inset following p. 44. University of Wisconsin Memorial Library.

This book too cannot escape the limitations Quetelet's long and productive life imposes on the historical researcher. The following is in some sense a biography but will pay almost no attention to Quetelet's family life. It is the story of the 'father of modern statistics' but will contain no equations. It is a history of a man whose most widely recognized 'discovery' today – the body mass index – was at the margin of his career and not popularized until a century after his death. It ignores much of his collaboration with Victorian statisticians and only briefly touches on his many projects to assemble a 'Global Physics', a tantalizing project that anticipated much of what today is grouped under the heading of climate science. For reasons explained below, the focus of the book will be limited to two interdependent aspects of his multi-faceted career: his creation of a controversial science of man – *physique sociale* – and his extraordinary work to create scientific institutions of observation at the same time.[7] Each, it will be argued, is impossible to imagine in isolation.

Initially, this book had been conceived only as an investigation into *physique sociale*, but after examining Quetelet's significant writings on the practice of science itself, it seemed clear that his *approach* to science was as valuable to understanding his career as his writings on social physics. Moreover, in examining the historical literature on Quetelet there appeared to be few efforts to examine *physique sociale* in connection with his various institutional roles: director of the Observatoire royal de Bruxelles, editor of the continental journal *Correspondance mathématique et physique*, permanent secretary of the Académie royale des science, des lettres et des beaux-arts de Belgique, state-sponsored traveler to observatories in France and Germany and correspondent to dozens of prominent nineteenth-century men of science. Yet as the sources of these various roles reveal, Quetelet's pursuit of large-scale projects for scientific investigation was not merely the means through which he accomplished theoretical research aims but, at times, appeared to function as ends in themselves. Though Quetelet did serious scientific research, it is hard to find an aspect of his career in which he was more successful, or one to which he was more dedicated, than in creating networks of scientific researchers. Though social physics is often the first thing mentioned in connection with Quetelet's career, this book suggests that even during its creation, it was only a part of his larger administrative concerns in Belgium.

Quetelet's plans for scientific research were directed from a number of influential posts in Belgium which connected him to a host of leading scientific figures throughout Europe. But they also required a particular kind of science worker, one that he admitted in *Sciences mathématique et physique* was very different from the natural philosophers of the seventeenth and eighteenth centuries. As seen in his proposals for institutions of observation and research, stated criteria for membership in a Belgian scientific elite and discussions with other *savants*, science worked best not in a laboratory or secluded study, but through the accumulation of large amounts of data from numerous observers. Because the new worker who helped to gather this information was expected to be standardized, interchangeable and to exhibit none of the extremes of genius or eccentricity, Quetelet's ideal social physicist appeared to be another average man called for by *physique sociale*. As Quetelet readily admitted, and desired, the new man of science would mark a significant break with the *savants* of the past.

An 'Intellectual Odyssey' of Nineteenth-Century Europe

The creation of two average men was the result of one extraordinary life. Though the project to count and analyse all the experiences of human behaviour may make Quetelet seem a real-life Gradgrind, dedicated to only 'facts and calculations', he was far from the sombre and taciturn parody found in Dickens.[8] Born into one of the most disastrous periods in the troubled history of the Catholic Netherlands, Quetelet used his enthusiasm and energy to help transform his country from a cultural and intellectual hinterland into a significant industrial and scientific power. Quetelet entered the world just a few years after the nadir of Belgian intellectual life in 1794 – when the occupying French closed the University of Louvain and transported the school's library to Paris – but left behind a nation that was one of the most advanced in Europe. During his lifetime, Quetelet either personally created or helped to build nearly every scientific institution in Belgium and the United Kingdom of the Netherlands,[9] including many of the important institutions today. In the process of doing so, he also found time to develop the foundations of modern statistics and quantitative sociology, as well as being among the leading researchers in the new sciences of meteorology and terrestrial magnetism. In the social sciences, Alain Desrosières has written that Quetelet's work was among the most important developments of the nineteenth century, suggesting that Quetelet's 'mode of reasoning … enjoy[ed] a posterity at least as important as that of more famous social thinkers', including 'Comte, Marx, Le Play [and] Tocqueville'.[10] Peter Buck has also claimed that 'as social scientists, it is with Adolphe Quetelet, and not seventeenth-century natural philosophers, that we share assumptions'.[11] It is certainly difficult to deny the truth of these statements in light of the current enthusiasm for Big Data and quantification in the social sciences and elsewhere, though Quetelet's 'mode of reasoning' should be construed broadly enough to incorporate his administrative work in standardizing scientific researchers alongside the ideas of *physique sociale*.[12]

There is another compelling reason to focus on these two aspects of his career at the expense of his more formal contributions to various disciplines: Quetelet's social physics and institution building put him at the centre of some of the most fascinating developments of nineteenth-century intellectual history and science. Attention to Quetelet's work and thought can engage the question of whether the revolutionary philosophies of the eighteenth century were subverted by the very mechanism these systems believed would facilitate progress: applying the tools of the natural sciences to the study of mankind. While many Enlightenment writers felt that science would confirm a world ruled by order and harmony and set out the conditions through which progress towards these goals could be made, by the end of the nineteenth century scientific investigation seemed to reveal chaos at best and a new form of progress through savage competition at worst.[13] Particularly troubling was the turn in the 'sciences of man', which by the time of Quetelet's death in 1874 had transformed into nothing short of an apology for European imperialism and retained none of the universalist assumptions that Quetelet had inherited from his *philosophe* heirs. Quetelet was the clear and obvious champion of the statistical sciences of man favoured by Condorcet, yet ended up as a principal inspiration for the eugenicists Francis Galton and Karl Pearson. In *Prophets of Paris* Frank Manuel famously called a similar transformation from egalitarian to competitive progress 'one of the crucial developments in modern intellectual history', and Quetelet's story is central to understanding this turn in the nineteenth century.[14] While Manuel did not include the Belgian, ending with Comte as his last prophet, the historian Lawrence Goldman has proposed that 'Quetelet's personal intellectual odyssey [was] a model of the early-nineteenth century determination to construct a "natural science of society"'.[15] This book largely supports Goldman's claim, but goes beyond the history of ideas to suggest that it was in the practice of science itself where the most important developments occurred in Quetelet's 'odyssey' through the nineteenth-century sciences of man.

Attention to the relationship between Quetelet's own science of man and his proposals for scientific practice can also help illuminate the dramatic changes in research practice during the century. Most importantly perhaps is how the boundary between science and other forms of intellectual activity influenced the production of scientific knowledge.[16] Certainly, once science had freed itself from intellectual (if not political) constraints, the increase in scientific knowledge was extraordinary as the nineteenth century saw tremendous advances in the understanding of heat transfer, electricity, magnetism, optics and evolutionary biology.[17] Was the flowering of the sciences directly due to the development of professional organizations, or was it because of what those organizations isolated scientists from, i.e., the moral constraints and concerns of writers, intellectuals, theologians and social theorists? Conversely, what was the consequence for theorists in the tradition of the Enlightenment who wanted to merge values, literature and science? Such questions can be addressed by studying a man who began work in the eclectic tradition of the French Enlightenment and ended up as one of the most successful scientific administrators in Europe.

If Quetelet is so important in two of the crucial issues of nineteenth-century historiography – the turn in the sciences of man and the institutionalization of science – why then is he relatively unknown today in spite of the tremendous growth of interest in the history of science in the past fifty years? Quetelet developed a massive network of correspondents throughout Europe including Villermé, Esquirol, Goethe, Humboldt, Laplace, Poisson, Gauss, Fourier, Arago, Bouvard, Malthus, Babbage, Herschel, Faraday, Maxwell, Forbes, Nightingale and Whewell. In America he was a regular correspondent with A. D. Bache and members of the American Philosophical Society; James Garfield praised his work on the floor of the United States House of Representatives. He was a member of over 100 professional organizations throughout Europe and the Americas and was responsible for founding not only the Brussels Observatory and countless other Belgian institutions but also the International Statistical Congress, which exists to this day. Yet outside of Belgium today his name is as rare as it was ubiquitous in the mid-nineteenth century, a time when his most popular work *Sur l'homme* was hailed 'as forming an epoch in the literary history of civilization'.[18] Why, despite a claim by George Sarton, the founder of the discipline of history of science, that *Sur l'homme* was 'one of the greatest books of the nineteenth century', has there not been a substantive work dedicated to Quetelet's life and thought since the early 1900s?[19] One possible answer is that Quetelet's long and successful career as a network and institution builder made him unsuitable to the two dominant trends of research in the history of science during the twentieth century: the old heroic narratives of the first half of the century and the post-war studies that sought either to complicate these narratives or create new heroes of unknown figures.[20]

Fortunately however, the discipline of the history of science has in recent decades begun to embrace scientific *praxis* as much as scientific ideas, and Quetelet has a claim as one of the more important enablers of scientific *practice* of the nineteenth century.[21] He played a crucial role in two of the century's most important scientific developments, both of which have somewhat confusingly been referred to as the 'second Scientific Revolution'.[22] Coined by Thomas Kuhn, the phrase was originally meant to encompass a series of changes of which 'one facet' was the vast increase in quantification. In the move towards quantification in the sciences, Quetelet was a participant at the end of a revolution, extending statistical techniques that had succeeded in other sciences into fledgling disciplines like meteorology and what would be known as the social sciences, including most notably his application of Gaussian distribution in astronomy to the study of social phenomena. At the same time, he and others helped create many of the statistical concepts that would feed James Clerk Maxwell's ideas on the development of molecular physics. The movement towards measuring with numbers began at the turn of the century with the mathematicians Laplace, Poisson and Fourier, three of the most important 'quantifiers' in Kuhn's story. Quetelet had in fact met all three of the quantifiers and it was just as Quetelet was creating *physique sociale* that Kuhn argued that the move from '*scientific law to scientific measurement*' was made.[23] As both theoretician and *propagandiste* Quetelet was one of the most prominent nineteenth-century savants in the move towards quantitative measurement.

Quetelet's role was even more noticeable in a different 'second Scientific Revolution', however, the one denoting the increasing professionalization of scientific practice.[24] Though considerable differences exist as to what professionalization was there is a general consensus that all sciences at the time were involved in developing forms of institutionalization to delineate their activities from other scientific and non-scientific activities.[25] As part of this programme, it was necessary to train and educate what Quetelet called a 'new class of men' to perform the duties of large scientific organizations.[26] As Quetelet claimed in two large volumes he published on the history of the sciences in Belgium, the transformation in the means of scientific production had profound consequences for the scientific worker.[27] The model man of science was no longer Voltaire's image of a solitary Newton divining the laws of the heavens, but rather the mass expeditions organized by the Académie française to determine the proper length of the metre. Quetelet believed that he had come of age at the end of an era of science driven by isolated geniuses and that future great discoveries would not be the work not of individual greatness, but of large-scale projects that would employ hundreds, if not thousands, of researchers. Quetelet's history and philosophy of science was not mere speculation, as he spent decades building institutions to conduct large research projects and a lifetime in Brussels teaching and training the men to fill the ranks of these groups. Though a movement of this size certainly had many participants, it would be difficult to find a more energetic and successful programme to organize European science along bureaucratic lines than what Quetelet conducted in Belgium between 1822 and 1835.[28]

Though Quetelet's embrace of what might be called bureaucratic science may not have the dramatic lure of Weber's 'disenchantment', Marx's dialectic of class struggle or Comte's grand stadialist account of historical development, this study of Quetelet's life and work takes seriously Desrosières's and Buck's suggestions above that Quetelet's ideas were among the most durable and lasting of the nineteenth century and that they may have more to tell us about the structure and organization of twenty-first century life and thought than much of the canon of nineteenth-century social thought. Yet it is only by seeing Quetelet's ideas, however inconsistent, alongside his practical vision of how large-scale institutions of science operated, that this importance can become apparent.

After Bielefeld

This change in focus represents one of several contributions this book hopes to make to recent studies of Quetelet. This current body of literature is a fragmented group that owes much of its confusion to two biographies written shortly after the turn of the last century.[29] Both books were concerned with placing Quetelet within a previously established narrative: for Frank Hankins it was the development of statistical techniques; for Joseph Lottin, the development of social theory. Both would now be considered unashamedly internal histories, an approach that had both its advantages and disadvantages. While each study situated Quetelet firmly within the confines of disciplinary history, they obscured the diversity of his thought and avoided what was most interesting about his career: the overlap in his thought and science that developed as he created both a quantitative science of human behaviour and the institutions to count these actions. Tellingly, both authors made only passing mention of *Correspondance mathématique et physique* and treated his administrative career in biographical gloss. Matters were not helped when Maurice Halbwachs, an important disciple of Durkheim, argued just a few years after Lottin and Hankins that there was no sense of unity in Quetelet's ideas. Just a generation after Quetelet's death, the noted sociologist wrote that 'the different parts that we distinguish in Quetelet's work are not connected, and ... each must be the object of independent discussion'.[30]

In the century since Hankins, Lottin and Halbwachs, however, there has not been a single-author monograph dedicated to Quetelet, a surprising omission considering the excellent work that has been accomplished in the history of statistics in the past thirty years. The impetus behind the explosion of works on statistics itself was a conference held in 1982 at the University of Bielefeld in Germany and the resulting two-volume study.[31] Aside from the impressive collection of papers presented during this conference, many of the participants went on to write ground-breaking studies of statistical development.[32] In the years between the early biographies and the Bielefeld conference, there were also occasional journal articles that sought to situate Quetelet in various disciplinary histories, the very eclecticism of which practically begged for clarity.[33] While these works varied drastically in terms of methodology and argument, they shared two common characteristics. The first was that the portrayal of Quetelet was limited by the relative lack of attention given to Quetelet's powerful position in Belgian institutions. At the least there was a lack of integration between his ideas and the institutions he served. The second feature of this literature was that, paradoxically, given the diversity of the works, they all relied almost solely on the biographies of Lottin and Hankins in presenting Quetelet's thought and career. While some attention was given to a few of Quetelet's most prominent works (mostly *Sur l'homme*), the interpretation offered by Lottin and Hankins was rarely expanded upon and the full range and impact of Quetelet's career was often missed. Collectively, these papers brought needed attention to the importance of understanding the history of quantitative thought, not only as a crucial development in thinking about numbers, but also the role the quantifiers had in the development of the 'hard' sciences themselves. Yet while all of this literature acknowledged the importance of Quetelet's ideas for understanding larger issues in probability and statistics, the understanding of the ideas themselves had hardly changed since the days of his first two biographies, published just four decades after Quetelet's death.

The failure of a fragmented approach was made clear in the publication of a series of talks during a colloquium held in Brussels on the bicentennial of Quetelet's birth.[34] Though the attention to discrete spheres of Quetelet's work did much to highlight knowledge in particular fields, the nature of the research made it difficult to say anything new about the man himself. Though each work provided a fine summary of one particular aspect of his career, there was no significant attempt at synthesis. Even when Quetelet's role as institution builder was recognized, there was no connection to social physics or any of Quetelet's other ideas.[35] There is of course nothing wrong with the approach of these essays if the goal was to understand the history of statistics, social theory or the creation of the Brussels Observatory, but they had the unfortunate side effect of forcing fragmentation on an interesting professional life.[36]

It might be said fairly that this book, through largely ignoring Quetelet's personal life, as well as his technical contributions to statistics and mathematics, similarly obscures the entirety of his work. It is also true that Quetelet's vast number of interests, writings and correspondents would make any one-volume biography necessarily incomplete. Yet in narrowing the focus to the relationship between Quetelet's *physique sociale* and his attempts to create institutions of science, it may be possible to recapture most clearly the attributes and ideas that made Quetelet such a figure of fascination and controversy in the nineteenth century. By doing so, it also may be possible to rethink some elements of nineteenth-century science and social thought. As should be obvious in the notes, but merits mentioning nonetheless, the task of providing this analysis has been made much easier because of the superlative work that has been done on Quetelet in the past three decades.

The 'Actualizer': Network Fields, Interaction Rituals and Friends

Because this book tries to situate Quetelet's project for social physics and institution building within the intellectual contexts of his time, a few words are necessary on the intellectual context in which *this* book has been written. It has been composed fortunately in the wake of a significant change in how the history of science has been written; fortunate, because it need not be consigned to categories of 'internal' or 'external' history, a dichotomy that many historians now find more problematic than useful in understanding the activities of the past.[37] Like anyone else, Quetelet drew inspiration, influence and ideas from both proximate and distal sources, and like anyone else, had his work both limited and expanded by friends, family, national self-conceptions and prevailing culturally and intellectually dominant trends. He was, to apply a designation that would have meant nothing to Quetelet but one he helped to make possible, highly socialized. So, however, were many other people in the same situation, none of whom ended up as the father of a science of man and the virtual creator of a national scientific identity. Therefore, the particulars of Quetelet's life and work must be treated with equal measure to the various contexts in which he lived and worked. The following paragraphs contain a review of some of the ideas that have led to this approach.

In navigating between the Scylla and Charybdis and of *text* and *context*, Quentin Skinner provided an early articulation of a third way of approaching the ideas of the past.[38] Though primarily concerned with the treatment of political ideas, Skinner's complaint that 'the writers of the past are simply praised or blamed according to how far they may seem to have aspired to the condition of being ourselves' seemed an apt description of how scientists and natural philosophers were treated for most of the twentieth century.[39] Even more so than in Skinner's world of political philosophy, early historians of science often 'foreshortened' the past, so that only those ideas that remained of contemporary importance were recognized. Furthermore, an excessive focus on the *context* in which the works were produced (whether they be national, institutional or cultural), Skinner argued, turned the writer himself into merely a mouthpiece for his age, reducing original and provocative thinkers to spokesmen for whichever important *Zeitgeist* the historian wished to investigate.[40] Skinner's solution – a complete historicization in which texts and contexts are only read with regard to what they meant at the time – might impinge upon the historian's raison d'être, but it remained a needed corrective nonetheless.

Though Skinner provided a way in which historians could reconstruct intellectual ideas through a close attention to textual *meaning*,[41] he did not offer a fully articulated structure of how ideas were actually transmitted across time and space. Nor is it obvious that men of science operated in the same way as political philosophers. This project, however, was adopted by sociologists in recent decades who have built up (it must be admitted) a rather rigid set of social causes for the production of scientific ideas. Foremost among them has been Pierre Bourdieu, who greatly expanded the Foucauldian project to recognize the importance of intellectual 'fields' in creating ideas.[42] Such fields consisted not only of the content of a particular discipline, but the relationship between the content and the way it was transmitted, debated and received by members of a given community. For example, the intellectual content of physics is the explanation of motion, but an articulation of the 'field' of physics is impossible without examining the particular world – the university, the corporate research and development laboratory, the conference in Las Vegas – in which physicists worked and the means through which they articulated their ideas. Though Bourdieu denied the sharp ruptures that marked Foucault's *épistèmes*, he retained the notion that it was impossible to examine a given idea without recognizing the interconnected set of social relations and material concerns within which that given idea was formulated. There could be no 'free-floating' ideas. Seeking his own reconciliation between the extremes of internal 'absolutist realism' and external 'historical relativism', Bourdieu posited a structure through which 'science' emerges through the competition of scientific workers over various professional and social resources.[43] Rather than ideas being derived from constant comparison of autonomous thoughts with an objective reality, success in science for Bourdieu depended primarily on how well individuals were able to exercise authority over a host of non-scientific factors.[44] Because no set of objective experiments could ever test the infinite number of possible scientific theories for a given phenomenon, consensus relied on bounding the possible. In an extension of Kuhn's articulation of paradigm resolution, Bourdieu argued that the ways in which boundaries are drawn must come from somewhere outside of the set of techniques considered as legitimate science. Science in this view became not the output of 'pure creators', but of 'actualizers who translate into action socially instituted potentialities'.[45] In evaluating Quetelet's contributions to the sciences, he might be best described as such an 'actualizer' rather than 'pure creator'. Though Quetelet was a gifted mathematician, his great successes (and failures) came about often through his ability to institutionalize ideas.

Since Bourdieu, the most extensive attempt to articulate a sociological theory of intellectual change has come from Randall Collins, who ventured to explain the history of thought since antiquity as the performance of the same scene over and over again on different historical stages.[46] It is a powerful argument with much explanatory potential but one that requires significant qualification if applied to the history of science, where his ideas have not had nearly the level of influence as Bourdieu's. In Collins's view, new ideas were almost exclusively the product of small, tightly knit and socially powerful intellectual networks. While it might make sense to see these groups as the ad hoc accumulations of influential and original thinkers, Collins seemingly put the cart before the horse, arguing that the social networks themselves produced individual genius: 'it is not individuals ... that produce ideas, but the flow of networks through individuals'.[47] Conversely, external (i.e., historical) explanations for new ideas were significantly underdetermined, since many people experience similar conditions yet few produce works of creative genius.[48]

In place of text and context, Collins inserted a series of local 'interaction rituals' through which certain groups gain influence in various societies. An important component in this view was personal contact, and Collins argued that the face-to-face exchange of social and cultural 'capital' was essential for intellectual transmission. In Quetelet's case, such direct encounters are crucial in understanding his work, as direct meetings with figures like Arago, Bouvard, Laplace, Gauss, Goethe and Humboldt, as well as less famous thinkers in Ghent and Brussels, played a profound role not only in introducing the young Belgian to new ideas but also to securing social capital in Quetelet's quest to convince governments to fund his many projects. In fact Quetelet relied as much on personal contacts as on scientific justifications to argue for new Belgian institutions. Collins's networks were all between five and seven actors – what he called 'The Law of Small Numbers' – so that capital could not be dispersed too broadly among a number of figures, and the expansion and contraction of such networks was, for Collins, what produced thought. In Quetelet's life, such numbers would vary drastically, but as he moved throughout various circles in Ghent, Brussels, Paris and the many observatories in Germany, he found himself often in such small and influential groups.

Though there is much to dislike about the idea of some sort of Hegelian accordion producing the most important ideas in Western thought, the theory of 'network ideas' flowing through an individual seems a remarkably apt description of Quetelet's intellectual and scientific life. Perhaps it would be incorrect to characterize many instances of scientific activity in these terms, but for Quetelet, an appreciation of his immediate intellectual context seems warranted. As a gregarious, motivated and charming person, Quetelet strived throughout his life to join and establish various networks of scientific thinkers; correspondence and personal meetings were in fact his chief modes of scientific activity. While it is the argument of this book that Quetelet himself was important in nineteenth-century thought, it was not necessarily in his creativity but in the force and success with which he 'actualized' the ideas around him.

While Collins's ideas were intended to explain only philosophies, Stephan Fuchs subsequently adapted the theory to scientific thought. Though the 'Law of Small Numbers' would not appear to apply to most scientific work since the mid-nineteenth century, historians of science have documented numerous cases where scientific ideas had come about in conjunction with the pursuit of social and cultural capital.[49] Furthermore, an explanation of nineteenth-century science as the product of social competition would seem at this point far more compelling than a story of individual genius or socially determined automatons. Indeed, Collins's theory even allowed for a bit of history! Though he specifically denied any 'external' account, he did allow that 'large-scale political and economic changes indirectly set off periods of intellectual change'.[50] Though Collins hid his allowance of outside forces behind the term 'indirectly', in this context the term was meaningless. Either history intruded on the autonomy of the 'network' or it did not; whether it was direct or indirect is a semantic game.

While the many ideas proposed by sociologists of science and thought are compelling, this book is intended as more than a case study on one particular theory of transhistorical action. As such, the motley jargon – fields, ritual networks, etc. – will be dropped in the body of the text.[51] However, Quetelet's story *was* largely the story of a succession of groups, and whether one calls these groups 'ritual networks' or 'friends', they were a vital element of the story. In all his dealings with friends, colleagues, administrators, famous scientists, observatory directors or journal contributors, Quetelet always displayed a great spirit and optimism, and there is little question that these social talents were among his greatest scientific talents.

Quetelet's personality matters not just as colourful detail but because it was crucial to the development of his average men. Bourdieu, Collins and Fuchs drew attention to how *individual* disposition to scientific practice could determine scientific production and therefore *personality* might make a better label than *habitus*. In other words, the particular talents of the leader of a scientific project – whether in conducting experiments, formulating new hypotheses, deducing laws or (in Quetelet's case) organizing large groups of observers – often indicated the parts of nature which the leader found valuable, and hence, structured his description of reality. This can easily be seen in gifted experimentalists like Galileo or Boyle, whose talents were in creating novel experiments and who, unsurprisingly, focused on experiment as the defining feature of science. The overlap of interest and discovery can be also be found in the case of Einstein, a deep thinker who was gifted neither in traditional mathematics nor in conducting experiments, but who believed science was best practised as a form of philosophy and whose conclusions often shocked the talented mathematicians and experimentalists of his day.[52] Much like the artist who argues for the innate superiority of the medium in which she most excels, the 'scientific method' often ends up reflecting the practice in which the scientific researcher feels most at ease.

Quetelet's talents did not include conducting novel experiments or, to be sure, deep philosophical thinking, but in using his charm and enthusiasm to bring together large groups of data and people in order to determine trends and laws in social behaviour. It should not be surprising after all that a man who excelled at directing large research operations suggested that science operated best through the accumulation of large amounts of data. Yet this connection between thought and practice has not been sufficiently made in any history of Quetelet. In fact Bourdieu, in a *hypothetical* description of one of his scientific actors, provides perhaps the best analysis of Quetelet's practice and thought:

> Those who come to head the large scientific organizations are obliged to impose a definition of research implying that the correct way to do science necessitates the use of the services of a large scientific bureaucracy – endowed with funds, advanced technical equipment, abundant personnel – and to institute as the universal and eternal methodology the survey of large random samples, the statistical analysis of data, and formalization of the results – in short, to set up the standard most favorable to their personal and institutional capacities as ... the yardstick of all scientific practice.[53]

As we will see, Quetelet did indeed insist on an 'eternal methodology' as well as 'abundant personnel', and his appreciation of 'large random samples' did relate to the kind of science in which he just happened to excel.

Though the book rejects the notion that the history of thought (scientific or otherwise) is the same performance of 'network actors' on different historical stages, it takes seriously the suggestion that the composition of a network, the competition that takes place within a given group and competition between groups, are vital parts of the story. In Quetelet's case we can see a succession of interactions with various networks, each of which would reflect and reinforce his evolving ideas. An excessive focus on outside 'forces' can simply be another form of historical determinism, however, so an attempt will be made to pursue three narrative strands: the historical conditions under which Quetelet worked, the composition of his immediate set of professional peers and the development of ideas in Quetelet's work.

An Overview

The multi-layered approach suggested above will be achieved through adjusting the level of abstraction throughout the book. The story of Quetelet's creation of two average men begins during a period described as 'The War of the Arts and Sciences' (Chapter One) when Quetelet was a young student deciding between careers in maths or literature. Belgian education was divided particularly by the new boundaries that were being created between science and art, and this chapter describes how Quetelet responded to the demands of his time. It is no secret that the creator of a statistical science of man read Condorcet and Laplace, but more surprising may be the young Belgian's appreciation for Romantic heroes like Chateaubriand and Germaine de Staël. Such a broad appreciation of art and science would not have been notable a century beforehand, but as this chapter demonstrates, the intellectual context in which Quetelet grew up was not well suited to continue such a broad course of study. To demonstrate the difficulties faced in the Low Countries in particular, the chapter concludes with the fleeting enthusiasm in the region for eclecticism and the possibilities offered by Victor Cousin's unique meld of art and science.

In Chapter Two, the focus moves in from the larger intellectual context to Quetelet and his close circle of friends in Ghent. Quetelet composed a large amount of poetry while simultaneously writing his dissertation in maths and often drew on his colleagues for support in the two projects. Yet not all of Quetelet's friends fared as successfully as he did, with several complaining about the conditions of employment in Belgian science. The eventual dissolution of the Ghent 'network' demonstrates that 'The War of the Arts and Sciences' was not simply about ideas debated by elites but an intellectual quarrel that had real consequences for young men interested in both mathematics and literature. This chapter finds evidence in Quetelet's poetry and opera librettos as well as in the troublesome outcomes for many of his friends for an increasing recognition of the strains in Belgian intellectual life brought about by the 'war'.

By 1819, when Quetelet left Ghent to take up a teaching position at the Athénée de Bruxelles, he had largely abandoned his plans to become a writer and had redirected his efforts towards making a career in the sciences. At age twenty-three, he had written next to nothing on statistics or probability, let alone the larger ideas of social physics that would be his most lasting legacy. As Chapter Three shows, Quetelet's new interests were sparked by new networks and institutions in which he could become an 'actualizer' of scientific ideas. Two institutions in particular helped in the creation of social physics and the average man of science: the Académie royale des sciences, des lettres, et des beaux-arts de Belgique and the Observatoire royal de Bruxelles. Quetelet was responsible for the creation of the latter and the complete reformation of the former. Based on his correspondence with colleagues, *éloges* (eulogies) for academy members and his lectures at the *athénée*, this chapter explores how Quetelet began to conceptualize the new science workers for his institutions at the same time he also imagined the best means to convince state administrators of the importance of these institutions.

The pursuit of an observatory led Quetelet to a world of nineteenth-century science far removed from his small group of friends. As seen in Chapter Four, in just over a decade Quetelet went from sharing a small house filled with poets and artists to meeting some of the most famous men of science of the era. In 1823 he travelled to Paris to begin research in the city's famous observatory, meeting Laplace and Poisson, but also the co-directors of the observatory, Arago and Bouvard. In 1829 he toured observatories throughout Germany, meeting Gauss, Goethe and Humboldt. In between the two trips, he helped found a continental journal called the *Correspondance mathématique et physique* that accumulated statistical accounts and other scientific notices from a wide variety of administrators and researchers. Again, his great talent for organization and enthusiasm served him well, as he made contacts throughout Europe that he used to support his projects back home. More importantly, however, this chapter shows that Quetelet spent as much time learning how to administer and direct large-scale institutions as he did learning about the particulars of the sciences of astronomy or statistics.

In Chapter Five, the narrative breaks to examine the creation of *l'homme moyen* and social physics between the years of 1827 and 1835, an intense period of time in which Quetelet did the vast majority of his work on statistics. Though the work was inspired in part by a genuine interest in reforming state institutions, this chapter demonstrates that *physique social* would not have been possible without both the cultural context of the 'war' and the immediate concerns of Belgium and the United Kingdom of the Netherlands discussed in the previous chapters. The Belgian Revolution of 1830 and the related delays in the construction of the observatory played a part, but so too did the numbers themselves. By this point Quetelet had already begun to formulate the *real* average men that he needed to staff the institutions of Belgian sciences, but the delay in constructing the observatory forced him to look elsewhere for data. He found them in the letters to the *Correspondance* which included tables on births, marriages, criminal behaviour, heights and weights. Though the form of *l'homme moyen* and *physique sociale* certainly had an intellectual lineage tracing back to

Condorcet and other enlightened ideas for a science of man, the chapter demonstrates that Quetelet's original conception of social physics was driven by the contributions of his network of associates built up over the previous decade.

In spite of the fact that *physique sociale* was created during a relatively brief period of time and that Quetelet quickly returned to large-scale data collection of natural, rather than social, phenomena after the observatory was completed, the tendency in the early biographies of Quetelet was to view social physics though the lens of ideas rather than practice. Quetelet was immediately criticized because of the determinist implications of social physics, and much of the subsequent discussion of the project has concerned the implications on free will for those who are the subjects of social physics. There is no doubt good fodder for such discussions: Quetelet was notoriously loose with his descriptions of how his various 'constant' and 'perturbing' forces acted, and his regular citations of Laplace did little to quell concerns that social physics was 'fatalist'. Yet as seen in Chapter Six, the focus on free will distracted many contemporaries from the more practical consequences for the actual practitioners of social physics, including their modern heirs in the quantitative natural and human sciences. Therefore, this chapter examines how Quetelet's social physics was first received and explains why this early criticism endured into the twentieth century. It concludes with a reconstruction of a debate organized by Quetelet in 1848 over the deterministic implications of social physics, a debate where one commenter – the future prime minister of Belgium Pierre de Decker – tried to defend his friend Quetelet. In doing so he first articulated the sense that there was *another* average man to be found in Quetelet's plans for a social physics.

The Conclusion returns to the particular Belgian context for the creation of the average men of social physics and Quetelet's history of Belgian science. *Sciences mathématique et physique chez les belges* admitted that the Low Countries could not keep up in the era when science was defined by 'men of genius', but provided a template for how the country could excel in an international world of scientific institutions. The new form of science would require *savants* spread out over the world, and Quetelet called the men of science who would fill these positions 'the New Argonauts'. Quetelet's dreams of a *physique sociale* were never fully realized in Belgium, but many of his collaborative ideas found full fruition in fields like terrestrial magnetism and meteorology, and Belgium took up an important place in European science and politics. In just one example, Quetelet's 1853 International Maritime Conference in Brussels is generally considered to be the first meeting where global temperatures and meteorological data were compared, an acknowledged forerunner of the Intergovernmental Panel on Climate Change.[54] In a brief epilogue it is suggested that the average men of these institutions may even offer a new way of understanding the legacy of the movement that Quetelet so believed he was following: the Enlightenment project to create a science of man as predictable as physics and astronomy.

Figure I.2: **Quetelet statue and Académie royale.** Quetelet's statue stands outside of the Académie royale de Belgique in Brussels, one of the many institutions of science he reformed or created in the 1820s. Much of his writing, from his poetry in the 1810s until his grand histories of Belgian science from the 1860s, is archived inside. One hundred yards to his left is the Palais Royale, where the kings and queens of Belgium still reside. Photo courtesy of the author.

Though the temptation exists to connect these New Argonauts to the modern bureaucrats found in Brussels, those housed in imposing modern steel and glass just a few Métro stops from Quetelet's statue on the lawn of the neo-classical Académie royale (see Figure I.2), such a connection would require a far different book from the one that follows. Quetelet's story told here – a young poet from Ghent who helped build the modern institutions of Belgian science, corresponded with the great scientists of Europe, created the first statistical science of man and envisioned a new of man of science – is, I hope, interesting enough.

1 LIFE IN THE WAR: THE END OF ENLIGHTENMENT IN BELGIUM, 1796–1823

A Civilization of Frontiers

Adolphe Quetelet witnessed an unsettling series of political regimes during his first twenty years in Flanders. Only two years prior to his birth in 1794, his boyhood home of Ghent was lost by the Hapsburg Empire to the newly formed French Republic. From there, the town was subject to the republican and then imperial programme of the modernizing French state, where the radical ideas of the French Enlightenment permeated down to the level of local administration. As an early twentieth-century guide to the region put it, 'French law and French administration laid the groundwork for a new social order' in Belgium.[1] While the exile of Napoleon brought a brief end to the conflicts, the Emperor's escape from Elba would not only send the restored Bourbon king to Quetelet's hometown to sit out the Hundred Days, but would also result in a fierce battle just fifty miles to the south, involving armies from all of Europe in the heartland of Belgium. As the great nationalist historian Henri Pirenne explained, these armies from Prussia, England and France left not just a physical impression on the landscape, but a cultural one as well. As it had been since the Treaty of Verdun in 843 AD, the Belgian identity, according to Pirenne, was forged by foreigners.[2] Napoleon's defeat at Waterloo guaranteed a stable regime for only another fifteen years – until Belgian independence in 1830 – but for a young Gantois mathematician and poet, the brief stability of the United Kingdom of the Netherlands (1815–30) must have seemed a relief.

The Catholic Low Countries of Quetelet's youth were not merely a physical location for state contestation. The region was also at the centre of an ideological struggle that defined political, social and cultural life in the first half of the nineteenth century in Europe. After the disruptions of the first two decades of the nineteenth century, politicians, scholars and writers alike tried to incorporate the radical political and social theories that were produced during the previous century in France as well as ward off their potentially disruptive consequences. The region was, in short, at the centre of a battle for the legacy of the Enlightenment.[3] Following the French Revolution, two powerful intellectual movements from neighbouring countries appeared in Belgium: the positivist scientific programme of the French *ideologues* and the Romantic critique begun in Germany. Here, as in so many areas, Pirenne saw a country that was 'half-German, half-French', and one whose intellectual and cultural life amounted to a 'civilization of frontiers'.[4]

In Ghent Quetelet was not only geographically equidistant between the two poles of Paris and Weimar, but intellectually divided as well, spending his youth studying his heroes Goethe and Schilling alongside the new forms of mathematics, physics and chemistry being taught in the Napoleonic Écoles. The son of a bronze sculptor, Anne-Françoise van de Velde, and a state bureaucrat, François Quetelet, the young Belgian was raised in a culture that encouraged a diverse array of interests. While the eighteenth-century *gens de lettres* and gentlemen of science had forged a consensus between art and science, the period of Quetelet's adolescence and early career was marked by a fierce rivalry of which Quetelet himself was hardly aware but whose consequences for his own ideas would be profound.

The debates between art and science that took place after the French Revolution have been identified by the historian and sociologist Johan Heilbron as disruptive to what he called the 'the intellectual hierarchy' of Enlightenment Europe.[5] In his view the early nineteenth century represented the time when 'men of letters lost their supremacy and ... the natural sciences came to occupy a dominant position'.[6] Though this statement is true enough for France, the Belgian experience requires some significant qualifications to understand the intellectual milieu in which Quetelet worked. The competing 'confluences' of thought (to use Pirenne's oft-deployed term) that influenced Quetelet, his friends and his university colleagues, were unique to the United Kingdom of the Netherlands. To understand the science of man Quetelet created, and the men of science needed to practice *physique sociale*, it requires an examination of a range of French quantitative thinkers and nineteenth-century Romantics that Quetelet often referenced. After examining the end of the Enlightenment consensus in Belgium, this chapter also explores the brief attempt to revive the harmony of arts and letters during the time of Belgian enthusiasm for the eclecticism of Victor Cousin. Though the movement ultimately failed, its reasons for failing anticipated some of Quetelet's later struggles and help explain why the Enlightenment consensus of arts and letters was unable to hold. Crucial to the development of both average men, Quetelet began to think about the practice of science and the demands of art at the very time when a 'war' had broken out between the two camps.

The War

In 1807 the counter-revolutionist Louis Gabriel Ambroise, Vicomte de Bonald (1754–1840), composed a small essay on a conflict that was tearing at the fabric of intellectual life in France. Never one to mince words, he called the work 'Sur la guerre des sciences et lettres'.[7] Bonald believed the war between 'science' and 'letters' had developed slowly during the previous century, but had entered a new phase as the two groups battled over influence in the triumphant Napoleonic regime and the subsequent direction of the *sciences morales*. While the moral sciences had traditionally 'ruled' through theology over the exact sciences and the arts, Bonald explained that the success of the Enlightenment claim to a universal and practical moral education had managed only a partial victory: while they discredited many of the positions of the Catholic Church, they were unable to replace them with a system of similar authoritative legitimacy. The authoritative and legitimizing *institution* of course existed in the person and bureaucratic reforms of Napoleon, but in Bonald's view it would be the *content* of this institutional message that was up for contestation.

The war would not be a fair fight. While the arts could rely on an 'ancient glory' of influence, their primary forms of 'argument' – epic poetry and high comedy – had been ruined during the previous century of artistic neglect by the *philosophes*. Bonald's political enemies bore the responsibility for this debasement of art because of their insistence on reducing all forms of expression, from architecture to music, to an imitation of nature.[8] The sciences, on the other hand, could benefit from the practical appeal of the *Encyclopaedia*, which had won many converts through its applications of simple technology to the pursuits of human happiness. Bonald had some notion of how his war would turn out when he wrote that 'all therefore announce the impending fall of the republic of letters, and the universal domination of the exact and natural sciences'. Yet presciently Bonald realized that the arts would not go away quietly. Anticipating the later polemical writings of Chateaubriand and Madame de Staël, he predicted that the arts would 'not die without glory and without vengeance: they will remain a strength even if they succumb in this war'.[9]

Bonald had perhaps the clearest perception of his self-described 'war'.[10] Writing in the aftermath of the socially and culturally disruptive French Revolution, Bonald was amazed at what he took to be a historically unprecedented discord between science and art. The ancients, he argued, certainly saw no difference between science and letters; for centuries, the two worlds worked together in the production of knowledge: science as the means by which people acquired knowledge and the arts and letters the way in which it was expressed and transmitted. For Bonald, the division of labour was clear and concise: 'science is the content [*fond*] and letters are the form'.[11] The complementary nature of science and letters progressed as such until its peak during the reign of Louis XIV, as 'the sciences became more literary and more ornate, and literature more knowledgeable'.[12]

The decline of this relationship occurred during the eighteenth century, and though Bonald argued that the *philosophes* had benefitted most from the consensus, they had also been the source of its disruption. Bonald blamed the schism on the same 'sophists' who abandoned classic forms of literary style and rhetoric for 'an appeal to invective and sarcasm'.[13] The excessive focus on form by the *philosophes* led to a relationship between art and science that had never been seen before. While there had previously existed content without form – knowledge without expression – the eighteenth century brought about the novelty of form without content – expression without knowledge. Lacking the thoughtful meditation of the great poets and writers before them, Enlightenment authors had simply chosen to 'pretty up a depraved knowledge with a pleasing form'. Bonald possibly could have tolerated the abandonment of serious thought had the form been better, but as he lamented, the eighteenth-century critics 'were not even [good] writers'.[14]

More surprising than his critique of the *philosophes* was Bonald's frustration with the technical language that was being developed in the sciences, specifically its reliance on numbers. The state of discord between art and science had reached the point that not only was form being delivered without content, but scientific production was being delivered in an incomprehensible form. Complaining that 'science speaks a language that is foreign to literature, and employs forms unknown to it', he echoed his fellow numerophobe Comte in his discomfort with the direction of the expression of scientific research. As counter-examples to the impenetrable language of the new scientific institutions, Bonald praised the example of Buffon, whose epic *Histoire naturelle* combined a deep well of knowledge and a 'pleasurable form' at different points. But even here, Bonald complained that Buffon could not do 'both at the same time'.[15]

An inability to be knowledgeable and eloquent had created a perverse consequence for how objects of inquiry were treated. Poems were being written on the sexes of plants while naturalists wrote of the 'affections of vegetables' and the 'families of shellfish'.[16] Conversely, humans had been reduced to just another species: 'It seems that one humanizes the material things in proportion to how one materializes man' Bonald insisted.[17] Through the combination of a devalued literary form and an increasingly technical content, the traditional relationship between content and form had been destroyed. Summing up the intellectual situation as the young Quetelet would encounter it, Bonald captured perfectly the state of the conflict: 'on the one hand there are letters without science; and on the other, science without letters'.[18]

The war between the arts and sciences was no less pronounced in the French-occupied Flanders of Quetelet's youth, where Ghent has been described as 'the natural intellectual centre' of the arguments that were just beginning to take place between French *ideologues* and German Romantics.[19] While the Enlightenment had had a strong influence on intellectual life in Flanders, especially during the time of the 'enlightened' rule of Joseph II (r.1765–90), a backlash against French ideas and a revived interest in German literary traditions took hold at the beginning of the nineteenth century.[20]

A landmark work in the transmission of German ideas, and one that Quetelet would cite in his early writings, was Madame de Staël's *De l'Allemagne* (1813). In the Germans, de Staël saw a 'spirit' and 'life force' that was absent in the 'sterility' of French literature, and she criticized harshly eighteenth-century philosophy, specifically the 'mocking heedlessness' of Voltaire and his 'imitators'.[21] Perhaps more surprisingly, given Quetelet's early praise for de Staël and his eventual career as an astronomer and mathematician, she contrasted favourably the German educational focus on language against the techniques of the natural sciences then being advanced in France and the Netherlands:

> The study of languages, which is the basis of instruction in Germany, is much more favourable to the progress of a child's faculties than those of mathematics or the physical sciences. Pascal ... himself recognized the defaults of minds formed first by mathematics: this study, in early ages, only exercises a mechanical intelligence: children occupying themselves only at this time with calculating, losing all the vigour of the imagination ... cannot acquire a truly transcendent mind.[22]

Quetelet admired Pascal and would have likely agreed with de Staël, but fortunately for the future of his 'transcendent mind', the primary schools of Ghent were not entirely devoted to 'mechanical intelligence' and included hallmarks of classical education like Latin and Greek. Quetelet certainly showed no lack of initiative or promise in mathematics, but it may be of note that at age twelve he finished first among his class in grammar but second in maths, receiving positive reviews and high grades in language, rhetoric and character alongside his grades in science and maths.[23] In his teens, just a few years before he would become a professor of mathematics and write his dissertation on conic sections, his instructor commented that his progress was 'higher in languages than in mathematics'.[24] Overall, the notes from his primary education describe a student proficient in the broad tradition of liberal arts, with little indication of a preference or talent for either side of Bonald's war. There was certainly no indication that the young Belgian would become 'the father of modern statistics'.

In spite of this broad education, Bonald's war may have had some influence on Quetelet's eventual career in the sciences. Though writing close to forty years after the fact, Quetelet explained the context of his education in an *éloge* devoted to the French astronomer François Arago, a friend who certainly lived a life that that crossed over the boundaries of the intellectual quarrel. Quoting the words of his colleague J. A. Berral, Quetelet reflected on the state of education in the Low Countries in 1815:

> A child, from the age of 13, having completed his fourth year, must choose between science and literature, and from them, to receive in one or the other a necessarily incomplete education; almost everyone wants to study science without having to do any studies in literature. It is a great misfortune for the younger generation ... We believe strongly that a man is great, even in the sciences, only when he has completed his literary studies.[25]

The 'great misfortune' of an 'incomplete education' Quetelet described was relatively new. For previous generations, especially among many of the writers that Quetelet admired, the choice between the arts and sciences was not necessary: little institutional or disciplinary specialization existed in the sciences, and the arts were still considered part of a broad intellectual spectrum. As has been recounted many times over, the professionalization of science and the creation of intellectual boundaries between science and non-science did not take place until the middle of the nineteenth century.[26] The irony is that this demarcation came about in large part due to the popularization efforts of many of the undisciplined and unbounded propagandists Quetelet admired and Bonald despised. The same thinkers who mixed humour, tragedy and history into their writings on science argued so strongly for scientific methodology that later eclectic thinkers were shut out of the new professional science. If Quetelet's generation was forced to 'choose between science and literature', it was in large part due to the calls of a generation that did not have to make this choice. Quetelet, arriving at least a few decades too late to be another d'Alembert or Condorcet, initially found himself in the middle of a war these thinkers had unwittingly provoked.

Before the War: Pascal, d'Alembert, Condorcet and Laplace

What kind of historical models might then have existed for a young student interested in probability theory, mathematical reasoning and the arts? In the decades and centuries before Quetelet, there was certainly no shortage of possible paths for a student interested in numbers, and the history of quantitative social thought in its early years includes a wide cast of true characters.[27] The intention here is not therefore an exhaustive account of the development of statistical models of social theory, but to present a series of sketches of thinkers who existed prior to the establishment of the strict disciplinary boundaries that Quetelet encountered in Belgium. Pascal, d'Alembert, Condorcet and Laplace were not only direct influences on Quetelet, but shared with him an interest in applying numbers to questions of politics, crime and human behaviour. While all could be said in hindsight to have inspired something like *physique sociale*, their ideas also represent a sampling of the forms of quantitative inquiry Quetelet could have followed had his career not begun in the midst of the war. By tracing the trajectory of these ideas, it is possible to see why Quetelet's eventual science of man looked so different from the kind of statistics imagined by quantitative thinkers writing before the nineteenth century.

Though different in temperament and style from Quetelet, Blaise Pascal (1623–62) was a model of the kind of diversity that had existed prior to the institutionalization of quantitative statistics. A child prodigy in physics, mathematics and mechanics, Pascal also composed essays on love, politics and religion. It is a mark of the lack of formal boundaries between scientific investigation and *belle-lettres* in the seventeenth century that some of Pascal's most influential comments on history, politics and philology appear in a work dedicated to proving the existence of a vacuum in a tube of mercury. In 'Préface sur le traité du vide', Pascal provided a distinction between two fields: those which can be improved and perfected because they are subject to 'reason' and 'experience' – physics, mathematics and medicine – and those which rely on the 'authority of books' – history, geography, jurisprudence, languages and theology.[28] Yet as Pascal's work demonstrated, there was no reason why a single individual could not explore both forms of knowledge, creating experiments to investigate physical phenomena while simultaneously interpreting classic texts to gain knowledge of God and man. Pascal's specific reasons for distinguishing between these types of knowledge in the 'Préface' were to discredit the ancient belief that 'nature abhors a vacuum' and to open a path for new knowledge. Because the possible existence of vacuums was a question to be investigated through experiment and demonstration, rather than consultation of canonical works, the appeal to ancient authority was in this case a category error. Empirical investigation was necessary at times, but it was not enough to capture all forms of knowledge.

A second important distinction made by Pascal was between two types of mind: *esprit géomètre* and *esprit finesse*. Usually noted because of its appearance in the *Pensées*, it is first mentioned in a short work entitled *Discours sur les passions de l'amour* (1652–3), where an attempt is made to explain the possibilities of finding love and beauty in a 'miserably short' life.[29] According to this work the geometric mind divides and classifies, reducing things to their smallest parts, while the sensitive (or delicate) mind appreciates things in their totality. Though Pascal certainly used the former kind of 'mind' in his investigation of nature, the geometric mind is described as 'slow, hard, and inflexible' and incapable of recognizing things in their entirety. On the other hand, the sensitive mind had a 'surplus of thoughts' which could appreciate many ideas simultaneously. Though a seminal figure in the history of probability theory, Pascal believed the two worked best while in tandem. This simple idea of the unity of all ways of knowledge had existed since antiquity, but by specifically separating and then reuniting *esprit géomètre* and *esprit finesse*, Pascal helped to create the conviviality between scientific and literary approaches to knowledge in the eighteenth century.

The young Belgian felt a strong affinity for Pascal; his eulogist Edouard Mailly wrote that Quetelet's favourite lecture at Ghent was on Pascal and that 'all his life he worshiped that great man'.[30] While no one would mistake Pascal as a supporter of the philosophical or political aims of eighteenth-century French *philosophes*, as a thinker interested in both mathematical reasoning and literature, he was an example of the successful era of intellectual interpenetration that would last until the end of the Enlightenment. His conception of a geometric and sensitive mind could also be extrapolated, with some care, to encompass the two dominant cultures of the arts and the sciences that Quetelet encountered.

If Pascal is too early and too religious a figure to be emblematic of the Enlightenment consensus among the arts and sciences among quantitative thinkers, then certainly Jean le Rond d'Alembert (1717–83) should do. Presented by the philosopher Ernst Cassirer as one of the leading thinkers of the later *philosophes* and heir to Voltaire, d'Alembert moved easily in salon life between mathematicians and writers.[31] As a mathematician, d'Alembert was most famous for his rejection of observational science in favour of a purely abstract and theoretical Newtonian model, creating in the process the intellectual space for the kind of law-like statistical accounts of society later explored by Condorcet and Quetelet. In doing so, d'Alembert also anticipated the encroaching empiricism and materialism that would come to represent for later critics the worst excesses of the eighteenth century. As a result, d'Alembert was rarely cited by the *ideologues* – despite a connection to Cabanis through Condorcet – and represented another path out of Enlightenment, one outside of the positivist commitment to a purely materialist explanation.

Though d'Alembert was influential in opening up the field of social mathematics, his career path was the opposite of that of later writers like Condorcet and Quetelet: he began his professional life as a mathematician but ended it as a writer. Though d'Alembert may have moved in the opposite direction, like Quetelet his journey was largely circumscribed by the political and intellectual context of his day. As the Encyclopaedists found themselves under harsh attack from the Jesuit *dévotes*, d'Alembert made a conscious decision to defend the *Encyclopaedia* and what he saw as the broad movement of the sciences and the arts that it represented. By being forced to defend this programme, however, d'Alembert had to abandon specialization as a simple mathematician. As the historian Thomas Hankins claimed, after the experience of defending his colleagues against the Jesuits, 'the transformation' away from specialization 'was complete'. After this moment, 'd'Alembert was a *philosophe* first and a mathematician afterwards'.[32] As a 'bridge between the professional scientists and the men of letters', d'Alembert took on the role of chief *philosophe* propagandist of science.[33]

Unlike many of his colleagues, d'Alembert seemed acutely *aware* of the realignment that was taking place in the intellectual hierarchy. As he put it in his 'Essai sur la société des gens de lettres et des grands' (1753), the truths of poets and artists were being replaced by the truths of scientific investigation, often ironically at the insistence of artists themselves. In an essay discussing the challenges faced by writers in currying the favour of political elites, he specifically contrasted the life of the *géomètre* and *homme de lettre*. Unlike Pascal's distinctions, where each form of thinking had its own role to play, d'Alembert's man of letters had to endure a constantly shifting series of opinions, and his fate was often determined by factors outside of his control. 'In matters of taste in *belle-lettres*', he wrote, 'esteem is always a little arbitrary, if not entirely so.'[34] In contrast, the *géomètre* could receive instant verification and legitimization of his theory through the success or failure of calculation. D'Alembert, pressing the knife further, claimed a poet would be 'useless' on a deserted island without anyone around to judge him, while the mathematician needed no such external acknowledgement. While the truth of this eighteenth-century theory of scientific production can certainly be questioned today, d'Alembert's formulation was an accurate reflection of many attitudes at the time. Though Voltaire and Diderot made no similar distinctions between the poet and the mathematician, with the latter strongly sceptical of *any* abstract calculation, other *philosophes* such as d'Alembert and Condorcet embraced quantitative reasoning and laid the groundwork for an overhaul of the intellectual hierarchy.

While few biographers or commentators on Quetelet have explored his links to earlier polymaths like Pascal and d'Alembert, preferring instead to look at either his sociological or statistical predecessors, the influence of the Marquis de Condorcet (1743–94) has been well established. In his definitive account of Condorcet's social thought, Keith Michael Baker claimed that the marquis' 'conception of a social science reached its full fruition in the social physics of Quetelet'.[35] While Quetelet was not born until two years after Condorcet met his tragic fate, he became acquainted with the work of the *philosophe* after meeting Condorcet's disciple Sylvestre Lacroix on a formative trip to Paris. Though Quetelet also met with Joseph Fourier, Poisson and Laplace at this time, Lacroix may have had the most direct influence on his statistical ideas: immediately upon returning to Brussels, Quetelet chose Lacroix's *Traité élémentaire du calcul des probabilités* (1816) as the model for his first course at the *athénée*. Though it has been argued that Quetelet differed from Condorcet in many of his assumptions, and even strict methodology, it is reasonable to assume that Condorcet's programme of a mathematical account of social behaviour helped shape Quetelet's later attempts at a statistical survey of social behaviour.[36]

While Condorcet was considered part of the broader Enlightenment movement because of his association with d'Alembert, he differed in important aspects from earlier *philosoph*es. In one famous version of this story, Frank Manuel's *The Prophets of Paris*, Condorcet functioned as the transitional figure between the broad-minded Enlightenment prophets and their more dogmatic successors

Saint-Simon and Comte.[37] The basic student–mentor relationships bear this out: prominent *philosophes* like d'Alembert and Turgot mentored Condorcet, and Condorcet in turn became the mentor of the arch *ideologue* Cabanis.

Though as an intellectual historian Manuel focused mainly on the internal progression of ideas among his prophets, research in the history of scientific institutions at the time provides a corresponding explanation for this progression. Famously shy, Condorcet's success was notably not to be found in the salons of his patron d'Alembert but in the 'proto-professional scientific organization' of the French Academy of Sciences, which he joined in 1768 and became president of eight years later.[38] As much as Condorcet's plan for a statistical accounting of the moral sciences opened a new field of research for Quetelet, his bureaucratic reforms and insistence on securing a comfortable place in government for the sciences helped contribute to the coming 'war' between the sciences and arts. In the salons, debate could move easily between art, drama, mathematics and chemistry, but in the academies as led by Condorcet and Turgot, the topics of debate were more narrowly focused. In creating a new field of research for Quetelet by continuing the Enlightenment search for a science of society, Condorcet also contributed to closing several more by moving the debate from the salons to the academy. As such, he may have not only been, as he was for Manuel, the pivotal figure in the transition from Enlightenment to positivism, but also the figure most representative of the transition from the Enlightenment consensus to the institutionalization of sciences that took place at the turn of the century.[39]

What Condorcet began, Pierre-Simon Laplace (1749–1827) completed. Famous for his injunction to 'apply to the sciences of mankind the methods of observation and calculation which has served us so well in the natural sciences', Laplace marked the beginning of an attempt at a statistical science of society completely divorced from the Enlightenment consensus.[40] Though a talented writer, Laplace was no dilettante, confining his investigation of social phenomena almost entirely to the language of statistical theory. He had met both d'Alembert and Condorcet, and owed his appointment at the French Academy to the latter, but no one would mistake him for a *philosophe*. Instead, as a mathematician and theorist, Laplace stressed the scientific and abstract justifications for a mathematical account of mankind more than the *Encyclopaedia*'s goal of alleviating pain and promoting happiness. While the Enlightenment authors saw science as the means to the eventual end of improvement in the material conditions of mankind, for Laplace a science of society was simply an extension of a successful methodological programme into a new field of research. Statistical investigation of society, as it was proposed in his *Essai philosophique sur probabilité* (1814), was an end in itself.[41]

Laplace also differed from d'Alembert and Condorcet in available outlets for publication and dissemination. While the latter two alternated between pamphlets and writing for the *Encyclopaedia*, Laplace's professional career overlapped with a more formalized form of scientific production; one, in fact, he helped to create. As Charles Gillispie described, Laplace's publications found their way into disciplinary journals that concretized the increasing specialization in scientific thought. Of this shift he writes: 'The day of the communication of scientific investigations through the medium of monographic research memoirs was, in any case, almost over'.[42] Quetelet's journal *Correspondance mathématique et physique* briefly reversed this trend in the 1820s through its incorporation of a broad range of disciplines, but the narrowing of scientific research fields had begun. By the time Quetelet met with Laplace in Paris in 1823, the leading work in probability theory and mathematics was not undertaken by intellectuals with broad research interests like Pascal and d'Alembert, or even Condorcet, but by highly trained and professionalized men of science. If Quetelet wanted to follow in the path of Laplace, it would not be in the manner of an Enlightenment *philosophe* but in the role of professional administrator of science.

The Romantic Critique: De Staël, Chateaubriand and Lamartine

At the same time Condorcet and Laplace were turning the investigation of man into what they believed to be a true science – creating a powerful symbiosis between the academies and government bureaucracy – artists and writers in France were awakening to their place in post-revolutionary politics. As they soon realized, there was not one. While Voltaire and Diderot had parlayed their popularizations of science and literary success into influential positions in the Prussian and Hapsburg courts, Napoleon was more likely to surround himself with men who limited their pursuits to science and engineering.[43] Though the Romantic author Chateaubriand had tried to appeal to the emperor as a possible liaison to the Catholic Church, his eventual exile proved how little writers had to offer the heads of state by the early nineteenth century. Left on the outside of the political and legal reforms in France, artists and writers instead looked to Germany for a new paradigm for artistic production. Realizing the truth of Bonald's belief that their outsider status was to be permanent, these authors created a radical new idea that left little room for a dialogue with the newly installed bureaucrats: art for art's sake.

Neither Quetelet nor any of his biographers mention a possible meeting between the father of French Romanticism and a young Belgian student busy writing operas and completing his dissertation. But it could have happened. In 1815, Louis XVIII had fled to Ghent following Napoleon's return, taking with him some of his most loyal servants and advisors, including Françoise-René, Viscount de Chateaubriand (1768–1848). As a contributor to the *Mercure de France* and the author of the passionate polemic *Génie du christianisme*, Chateaubriand was a popular figure in Catholic Flanders. Though he had the ear of the exiled Bourbon king, he preferred to stay away from the 'petty intrigues' of the court during his time in

Ghent, sequestering himself in a convent he appreciated for its 'calm demeanour'. It may have been a characteristic overstatement by Chateaubriand to claim that his 'passage through [Ghent] was announced as that of a missionary or doctor', but it is unlikely that the nineteen-year-old Quetelet could have missed his presence.[44]

Even without a physical meeting, the influence of Chateaubriand on the young Quetelet was immense. Writing in his *Essai sur la romance* in 1823, Quetelet ascribed to Christianity the same importance in the development of drama, art and poetry that Chateaubriand first examined in *Génie du christianisme*.[45] Additionally, as seen in the next chapter, Quetelet's poems and operas are filled with heroes who embody the concept of the tortured artist that Chateaubriand developed in *Mémoires d'outre-tombe*. Along with de Staël, Chateaubriand would be the aesthetic guide to most of Quetelet's Romantic writings and his introduction to the literary philosophy of the German territories.

For a future statistician and observatory director like Quetelet – one who would later claim for perfection the statistical mean of physical and moral characteristics – the great French exponent of German Romanticism makes for an odd influence. Not only was Chateaubriand's political conservatism foreign to Quetelet, but the French Romantic was at his most vicious in attacking the element of the eighteenth-century thought that Quetelet was most interested in: scientific and naturalized descriptions of social behaviour. Chateaubriand made his criticisms quite clear in his first major essay, *Essai sur les revolutions* (1797), claiming that the social philosophy as expounded by the *Encyclopaedia* was nothing more than a 'sophistry' of 'vain systems'.[46] In a broad comparative work that contrasted the ancient revolutions of Athens, Egypt and Rome with the modern one in France, Chateaubriand blamed the ensuing social disruptions on the *philosophes* for their harsh treatment of Christianity and ironic distance towards affairs of the state. While the philosophers of antiquity took great pride in inserting themselves into politics and government, the recent philosophers abstracted themselves from their concerns: 'Shut away in their office [*cabinet*], they spend their mornings on books about a war where they have never been, on a government where they will never take part, and a natural man that they have never studied'.[47]

While his first essay was motivated by anger at the Enlightenment for what he believed was the destruction of the Christian religion, by the time of *Génie du christianisme* (1802), Chateaubriand had refined his critique of the eighteenth-century philosophers. His defence of the Christian faith was founded on the need to preserve a sense of mystery in the world against the classifications and 'dissections' of the *philosophes*. This mystery, he believed, had flourished in the Middle Ages, between the persecutions of Christians in the Roman era and the most recent attacks of the eighteenth century, specifically by the 'Babel of science and reason', the *Encyclopaedia*.[48] Early chapters were dedicated to proving the importance of the Christian faith in every conceivable form of art and morality, an importance overlooked by eighteenth-century critiques of the Church.

Chateaubriand's most forceful attack would be reserved for the Enlightenment promotion of the sciences, however, which he felt had unnecessarily isolated science from other disciplines, including literature and theology. Though the professionalization of the French Academy as created by Turgot, Condorcet and Laplace was a recent phenomenon, Chateaubriand was keenly aware of the new organizations, classifications and terminologies that were being created to distance scientific research from other forms of intellectual activity. To be sure, he was sceptical of such claims. If the great scientific investigators of the past had been Christians and *gens des lettres*, Chateaubriand asked, why now had the investigation of nature been divorced from the investigation of God? He compared Bacon, Newton, Leibniz and Pascal favourably to the 'childish' writings of Condillac and d'Holbach, writing that 'in general, the inventors of geometry have been religious'.[49] Furthermore, such 'great men' as the seventeenth-century mathematicians were 'very rare' and the average geometer working in their shadow was 'condemned to a sad obscurity'. The reason, simply, was that the vast majority of mathematicians and scientific investigators had little to offer that was novel. Chateaubriand was unequivocal in placing the work of art above that of mathematics:

> At all times, every country offers the same example. The mathematicians will cease therefore to complain if the people place letters before sciences! The man who has left a single moral precept, a lone feeling left to the earth, is more useful to society than the geometer who has discovered the most beautiful properties of the triangle.[50]

In a sense, Chateaubriand's formulation inverted d'Alembert's lament that external justification limited the appreciation of excellence in letters. Instead of d'Alembert's stranded poets on a deserted island, it was the geometers who suffered the fate of oblivion.

In claiming the superiority of artists, Chateaubriand was faced with a difficult question: if artists 'were more useful' to a society, why had the sciences seen such a growth in popularity over the previous century? For Chateaubriand, it was not a result of their technical capacity or the truths uncovered, but the relative simplicity of science. Geometers, chemists and natural historians had simplified nature to a degree that was easy to understand in contrast to Christianity's world of mystery and unanswerable questions. As he wrote disapprovingly, 'the law of gravity can be known in the lowest schools; and even a child can scribble the figures of geometry'. In this view, scientific practitioners had lowered the standards for what counted as knowledge of the world by reducing it to such a degree that it was understood by everyone; classification and reduction diminished understanding. Chateaubriand even evoked a Kantian scepticism of the limits of empirical investigation, writing that 'the brevity of our life, the weakness of our senses, and the crudeness of our instruments ... are opposed to the general formula [of life] that God forever hides from us'. The sciences at the time may have been able to offer some descriptive accounts of nature but could never create knowledge like art and religion. In short, said Chateaubriand of the intellectual hierarchy at the time, 'our sciences can *decompose* and *recompose*, but they cannot *compose*.'[51]

This is not to say that Chateaubriand saw no purpose for scientific investigation or any value in the discoveries of the eighteenth century. Instead he believed that like art, poetry and *belle-lettres*, science needed to be subservient to a moral philosophy to be relevant to individual lives. Though different in political philosophy and social outlook from the *ideologues*, Chateaubriand shared with the positivists Saint-Simon and Comte the belief that the specialization was leading to a world in which knowledge was being lost, rather than accumulated. Without a guiding philosophy – Christianity for Chateaubriand, sociology for Comte – the discoveries of the newly professionalized sciences could be no more than mere curiosities. Succinctly characterizing the belief of many at the time who wanted an instrumental, rather than philosophical, science, he finished his chapter on 'Chimie et histoire naturelle' in *Génie du christianisme* with a call for unity between the sciences and the arts: 'We conclude that the defect of our day is to separate too much abstract studies from literary studies ... to sacrifice that part which loves to that which reasons.'[52] Ironically, due in large part to Chateaubriand's influence over future writers, the separation of 'abstract studies' and 'literary studies' only grew.

As noble as Chateaubriand's sentiments might have been for their attempt to bring a truce to the 'war', he was hardly a neutral observer. An indication of the rigidity of his position is found in an early review of the writings of Madame de Staël (1766–1817), his eventual partner in the creation and promotion of the Romantic ideal in France. As an early admirer of the *philosophe* programme, de Staël's migration from the Enlightenment camp to the side of Chateaubriand indicated how little room there was for compromise at the beginning of the nineteenth century. Initially, de Staël was so strong a believer in the eighteenth-century form of progress that one contemporary described her as 'the widow of Condorcet', and her first publication was viciously attacked in the counter-revolutionary journals *Mercure de France* and *Journal des debates*.[53] But after a period of exile and a visit to Germany, de Staël became a reliable warrior for the side of art against what she saw as the excesses of scientific reasoning. Her intellectual development was indicative of the challenges faced by any progressive thinker at the time, and Quetelet knew her work and story as well as any other French thinker of the era.

The book which first inspired the ire of reactionary conservatives was *De la littérature considérée dans ses rapports avec les institutions sociales* (1800). Though it did not have nearly the influence or importance as her 1813 *De l'Allemagne*, *De la littérature* was a powerful presentation of the Enlightenment conception of progress and the importance of scientific reasoning. Citing Condorcet, with whom she shared a vision of indefinite progress towards perfectibility, she claimed that 'the sciences have an intimate connection with all of the moral and political ideas of nations'.[54] Reviewing the book, Chateaubriand and the critics at *Mercure de France* may have been able to tolerate this 'connection' between science and morality, but de Staël went further to describe science as the engine that drove *all* forms of knowledge: 'I say even more, the progress of the sciences is necessary for the progress of morality'.[55] Necessary? This was going too far. While the investigation of nature through classification, dissection and experiment may give some account of nature, and therefore help contribute to man's understanding of his own morality and God's creation, Chateaubriand could not tolerate a morality that was subservient to science. In de Staël's conception of the intellectual hierarchy, literature itself took the place of religion as the model to which all other forms of knowledge must conform.

It is not hard to see why de Staël's initial conception of a scientifically-achieved morality and her wholesale adoption of Condorcet's optimism was so palatable to a young writer and mathematician located in a country bordering France and Germany. Quetelet's later investigations would show almost a complete reliance on the importance of the *means* of scientific investigation – data collection, institutional organization and methodological standardization – rather than on the *ends* of morality. Even the subjects of Quetelet's later research into society, such as reducing criminal acts, were seen as much as evils of inefficiency as of morality. Chateaubriand, on the other hand, cared little for *how* science operated so long as it confirmed the morality taught in the Gospels. De Staël, by combining an appreciation of the Romantic poets of the Middle Ages and a teleological view of progress through science, provided Quetelet with a model that allowed the combination of the two dominant interests of his youth.

The idea of Condorcetian progress with literature substituting for empirical and rational science did not last long. Almost immediately after *De la littérature* appeared, Louis de Fontanes published two scathing attacks in his conservative journal *Mercure de France*, specifically targeting de Staël for her supposedly naïve belief in progress and perfectibility. The attacks were successful. Ten years after *De la literature*, de Staël abandoned most of her progressive language and turned to complaints over 'the infallible march of calculus'. Aside from her objections to the overabundance of mathematical training in schools seen earlier in this chapter, where she claimed that too much maths led to a 'mechanical intelligence' and the loss of 'all the vigour of the imagination', she now echoed Chateaubriand in claiming a limited audience for abstract reasoning: 'nothing is less applicable to life than a mathematical reasoning', she wrote. For her, mathematical reasoning led to only true or false answers, a dichotomy unsuited to a 'life more complicated'.[56]

Notable for its possible impact on Quetelet, who often referenced *De l'Allemagne*, was de Staël's allowance for 'probable truths … [which] can serve as a guide in affairs, as they do in the arts and in society'.[57] Though Quetelet's eventual theory of *l'homme moyen* was roundly criticized for being deterministic, his averages and probabilistic models introduced nuance lacking in Condillac's statue-man, Condorcet's sketch or Comte's stadialism. Though de Staël was likely unaware of the revolution in probabilistic thinking at the time, her belief was consistent with what Lorraine Daston has described as 'classical' probabilistic thinking.[58] Indeed, it would be just these probable truths that would take up much of the 'mathematical reasoning' of the nineteenth century, a movement motivated in large part by Quetelet. However, in criticizing mathematics in 1810, de Staël was much more likely to be thinking of the 'force' determinism of mathematicians like Laplace and the social programme of Cabanis than the more nuanced attempts at a social mathematics that followed. Her criticisms and retreat clearly won over her former detractors at the *Mercure de France*; after a second edition of *De la littérature* in which she acknowledged the journal's critiques, Fontanes promised to never again criticize her in his publication.[59]

Within the span of a decade, de Staël went from being 'the widow of Condorcet' to a favourite of Chateaubriand. Part of this was due to disillusionment with the progress of Napoleon's version of the revolution. Exiled in 1803 for her novel *Delphine*, de Staël's experience with the new literature in Germany certainly helped temper her enthusiasm for the Empire and the rationalism it was supposedly spreading to the rest of Europe. Yet the most compelling reason for her change in thought was likely the intensification and radicalization of the two intellectual camps of positivism and Romanticism; she may in fact have been the first casualty of Bonald's war. At the end of the eighteenth century in France, it may have been possible to hold to Condorcet's view of progress while still declaring literature to be the preeminent means of acquiring knowledge about the world, but as the attacks by Fontanes and Chateaubriand demonstrated, an ecumenical approach was open to harsh criticisms by at least the time of the creation of the Empire; there could be progress, and there could be the arts, but there could not be both. While de Staël may have felt the sting from the criticism of her fellow *gens* (and *femmes*) *de lettres*, the new standardization of the sciences in France at the time, to say nothing of their complete subservience to Napoleon's bureaucracy, made her break from the Enlightenment that much easier. As her experience in the 'war' demonstrated, there were significant consequences for anyone daring enough to straddle the line between literary and scientific ideas.

By the 1830s the war between the arts and sciences was nearly over. The French poet Alphonse de Lamartine (1790–1869), a contemporary of Quetelet who praised Chateaubriand and de Staël as 'two lively protesters against the oppression of the soul and the heart', was the most vocal warrior of this era. In both tone and content, Lamartine's work makes clear that he was writing in the *aftermath* of a conflict and not during one of its formative battles. Concerned about the 'insolent tyranny of their triumph', he accused mathematicians and *savants* of trying to 'condemn and destroy ... all the moral, divine, and melodious parts of human thought'.[60] He complained of lacking even intelligent men with whom to converse about great works of literature: 'no one understands them ... I read them to myself some days'.[61] Lamartine may have been overstating the point, but his complete frustration was itself evidence of the resolution of the war and of the limited state-sponsored opportunities available to someone with an interest in poetry by the middle of the century. These, however, would not be Quetelet's concerns; by the 1830s he had already sufficiently insulated himself within the very institutions Lamartine decried.

It does not appear that Quetelet speculated much on Lamartine or even knew his name – his interests in poetry reduced to pleasant openings in letters to old friends by the 1830s – but it would have been interesting to know his reaction to a fellow admirer of de Staël complaining of quantitative reasoning as an 'abhorrent number, that negation of thought'. While Lamartine's essay in 1834 went on to describe how a new art may emerge – one which treats man 'as sincere and in its entirety' – he shows no interest in a poetry reconciled to the sciences of the time. Speaking of this new form of poetry, he claimed it 'does not want to be a puppet (*mannequin*) any longer, nor a machine'. Instead he viewed poetry in direct opposition to the sciences. Just a few decades after Chateaubriand's warnings, Lamartine confirmed the worst fears of the early French Romantics, claiming that 'The first thing that I have in mind now is for [poetry] to examine the puppet, and to show ... poetry alone in poetic works'.[62] By strongly rejecting the influence of science or 'the machine' in conceiving poetry and the arts, Lamartine argued for an artistic life every bit as isolated and self-referential as the scientific objectivists at the British Academy and with self-confidence equal to Laplace. Furthermore, by explicitly linking his idea of a self-made and self-determined art to the most powerful movement in the arts – Romanticism – Lamartine took artists and writers further away from the sciences, siphoning off the last possible hope for a new consensus. Chateaubriand and de Staël may have initially wanted to bring science into accord with a more broad-based programme of religious and moral instructions, but their heirs like Lamartine had no such illusions about the newfound autonomy of science.

Eclecticism in Belgium: A New Consensus?

If the war of sciences and letters could not be resolved in France or Germany, the 'civilization of frontiers' in Belgium offered an exciting but ultimately failed path to peace: eclecticism. Created by the French philosopher Victor Cousin, eclecticism might have been the perfect fit for a country that found itself between two competing philosophical systems. At the least it could have served as a guide for a poet and mathematician brought up on Romantic ideas and the progressive dreams of the *philosophes*.⁶³ The seeds of this philosophy in fact grew out of a response from some of Quetelet's friends and colleagues. Belgian philosophers in Liège led by Pierre Rousseau justified their interest in eclecticism in order to resist the 'invasion of the French *philosophes*' and 'Voltairisation' in Belgium during the second half of the eighteenth century. It was not until the renovation of the Hapsburg educational system in 1815 by William I, however, that the movement gained serious traction in the universities. This 'new era' was symbolized by the appointment of Quetelet's close friend the Baron de Reiffenberg at Louvain, whom the early Belgian intellectual historian Maurice de Wulf called 'the most significant philosopher of the university group'.⁶⁴ Quetelet worked in the 1820s during a period when some of his closest colleagues favoured eclecticism, and during a time when he was trying to balance literary and scientific work. If there was any coherent philosophy available to sort out his conflicting interests, eclecticism might have been it. Yet the ultimate failure of a system of thought meant to balance the arts and sciences points to the larger dilemma which faced Belgian thinkers, writers and *savants* in the early nineteenth century. Trying to subsume the sciences within a larger moral and religious framework, eclecticism had little success in influencing Belgian institutions of science.

De Wulf identified Quetelet's friend Reiffenberg as part of a 'triumvirate' of eclectic thinkers centred in Louvain along with Sylvain Van de Weyer, a fellow professor, and Pierre François van Meenen (1772–1858), a lawyer thirty years their senior and the inspirational leader of the group. Van Meenen in particular laid the groundwork for eclecticism in Belgium, beginning his early studies during the final years of the Hapsburg rule when the doctrines of the Enlightenment were at their strongest in Belgium. Van Meenen's primary philosophical target was Condillac's sensationalism, which though a refined version of the 'brutal materialism of the Encyclopaedists and revolutionaries', was still unsatisfactory to a group who endeavoured to retain concepts like the soul and free will. Although he agreed with Condillac that the physical body was passive, van Meenen held to a Cartesian dualism that allowed for a 'passive-active' soul that could influence the body as easily as sensations (or God). Furthermore, he believed that it was the role of philosophy to describe the resulting actions of a body as it related to the impetuses of soul, nature and God. As such, the philosophy of the eclectics was necessarily ad hoc, or, as van Meenen put it, 'humanity speaks and philosophy listens'.⁶⁵

Without an academic position and in the absence of any institutional support, van Meenen would have to wait another generation before being adopted in the universities by Van de Weyer and Reiffenberg, and until the late 1820s to link up to the popular and influential writings of Victor Cousin. By 1840, a small pamphlet had been published which served as a sort of manifesto for this group, consisting of three lectures given between 1827 and 1830 by Van de Weyer, Reiffenberg and Cousin. In the introduction the editor proposed a philosophy of simplicity and clear concepts in the style of the 'Scottish school' and the 'French eclectics'.[66] This statement of principles in *Collection d'opuscules philosophique et littéraires* is valuable not only because of the geographical and chronological overlap between it and Quetelet's most enduring works, but because Quetelet knew and referenced all three authors: Reiffenberg was a close friend, Van de Weyer a fellow member of the Académie royale in Brussels and Cousin was referenced often in Quetelet's signature work *Sur l'homme*. Though Quetelet had been introduced to the writings of the Romantics after reading de Staël and his work bore the methodological stamp of Enlightenment mathematicians, eclecticism offered a third way to incorporate such disparate approaches at a time when the traditional consensus of the previous century was on the decline.

The *Collection* authors focused primarily on social and political concerns. Though schooled in the formal philosophies of the ancients and seventeenth-century rationalists, the driving motivation was to establish a philosophy that would stem the revolutionary excess and wars that had spilled over from France into Belgium. The Baron de Reiffenberg (1795–1850) was unequivocal in finding a culprit:

> The majority of the changes in the social order were uniquely linked to the spirit of immorality, dizziness, disorder and atheism which accompanied [the *philosophes*]. In appearing as the necessary cause, they reserved for themselves the title of 'philosophy' and struck down that which is truly meant by this name.[67]

For the baron, true philosophy instead was that which treated man and his relationship to nature in its complete entirety, without subdivision or dissection and without experiment or investigation of fragments. The *philosophes*' failure to do this had led to 'immorality, dizziness, disorder, and atheism'. The eighteenth century, despite claims to a universalism, in his view had only looked at a 'mutilated thought' broken into many pieces that had little relationship to the 'entire man'. Man as whole, to say nothing of society, had an infinite number of possible departures for investigation, and the excessive focus on one specific element meant an abandonment of the search for true knowledge of mankind's condition. A true philosophy, which eclecticism endeavoured to become, was that which 'contemplated man with an infinite multitude of views'.[68] The *philosophes*, by contrast, obliterated this complete man from their system and opened the door for what Reiffenberg called the 'hideous procession of devastation and death' that took place around the turn of the century.[69]

Though pleased that the philosophy of the Encyclopaedists and Robespierre (Reiffenberg made no distinction between the two) was on the wane in Belgium, the baron saw a new threat in their materialist ancestors in France. Positivism, though yet to receive much support outside of a few small societies in Paris, was a problem for a true philosophy for two reasons: it intentionally avoided answering 'why' questions and, for many, was incomprehensible. On the latter point, Reiffenberg summed up a view likely held by many upon first encountering Comte: 'Is this method a mystery reserved for only a few of the initiated? Why does it hide itself in an inexplicable labyrinth of complicated subdivisions? Why is it based on a painful and barbaric terminology?'[70] Most disturbing was Comte's decision to limit himself to mechanistic explanations. To focus only on *how* things operated without inquiring as to *why* they operated was not only illogical; it also invited a dangerous materialism. Additionally, without understanding the first principles of the universe, positivism simply examined how something worked, without any idea of what that *something* was in the first place. Its 'discoveries' therefore could not be considered knowledge, meaning it was neither a philosophy nor even a science, but only an enclosed and self-referential set of methodologies and practices unrelated to the outside world.

In what must have been a reaction not only to the political currents but also the changing Belgian landscape, Reiffenberg spelled out in detail the consequences of a self-contained materialism that represented 'the universe as a vast factory operated solely by a steam engine'. The simile was well chosen, especially for Belgium in the early nineteenth-century, as Reiffenberg was concerned with the expansion of an 'industrialism which seeks to dethrone morals and replace them with a financial aristocracy'.[71] Because the industrial changes stemming from this intellectual movement were already being made manifest, the baron offered a comprehensive overhaul of the educational system which included an emphasis on moral education over training in the sciences, a plan directly in opposition to the French model of specialization favoured by the United Kingdom of the Netherlands. Though his friend Quetelet ultimately proved successful in dealing with the path described by his friend, Reiffenberg's predictions had not been far off the mark.[72]

While Reiffenberg wanted a guiding philosophy for a pedagogical system equivalent to broad instruction in the liberal arts, the second author in the *Collections*, Sylvain Van de Weyer (1802–74), did the most to extend eclecticism as philosophy proper. In 'Discourse sur l'histoire de la philosophie', Van de Weyer specifically presented eclecticism in Belgium as an alternative to the German and French systems: the former he called a 'vapid idealism' removed from true investigation, the latter an abuse of philosophy. Apparently believing that the faults of the Germans needed no further explanation, Van de Weyer focused his critique specifically on the destruction of 'true philosophy' during the eighteenth century in France. That era, though producing a number of excellent 'physicians, chemists, naturalists, and industrialists', had turned its back on the very philosophy that had inspired it in the first place, and abandoned the most important aspects of knowledge which 'escape their scalpel or their microscope'.[73]

Van de Weyer linked a philosophy of an eclectic *bricolage* to an appreciation of the various manifestations of God, believing that the further science moved away from knowledge of the divine, the more it became merely instrumental. The eighteenth century had proved that a completely scientific account of nature and man left only atheism and a 'distressing spectacle' of every man out for himself. By subdividing man and nature into discrete concepts, the sciences had lost sight of the whole man, which was the proper study of investigation. A man trying to understand the meaning of life through science, Van de Weyer believed, was like a man trying to understand the beauty of a great poem through an analysis of each individual word and letter: 'these things separated by themselves are without life'. Instead of the 'cadaver' that the *philosophes* wanted to investigate, 'true philosophy' taught men what they had in common and led to a purpose of belief. Not only did this 'purpose' make man whole again, but connected him to others through shared and universal beliefs. 'Philosophy is in some sort placed outside of the sciences', Van de Weyer wrote, and the goal of eclecticism was to then find that philosophy.[74]

Van de Weyer did however admit that the Enlightenment critique of metaphysics had been forceful and complete. There could be no going back to Platonic ideal forms, the monad or the cosmology of the medieval Church; science had proved certain truths that had to be accepted. Therefore, because no previous system of belief could be adopted completely, it was the responsibility of the eclectic philosopher to find and combine the various truths found in Plato, Aristotle, the Church, Descartes and Leibniz into a coherent picture of man's place in the universe. Each of these philosophical traditions, it was true, had been discredited in some manner by the investigations of the eighteenth century, but viewed *en ensemble* they could give a true and accurate picture of the world. While the sciences could break down nature into smaller and smaller parts for study, eclectic philosophy was needed to look at the most important questions: what is the good life? What is the meaning of existence? What should we do?

This programme, as should be clear, was not exactly philosophy per se but more of an applied history of philosophy, where the canonical authors of the past could be harvested for remaining truths and then repackaged as moral education. In a sense, the approach was similar to the Enlightenment authors whom eclecticism so often criticized, but instead of simply mocking the inconsistencies and failures of philosophies throughout the historical record like Voltaire, the goal was to go back and revive whatever remained of them after all of the inconsistencies and contradictions were removed. As both Reiffenberg and Van de Weyer made clear, the available philosophies open to plundering were not limited to the ancients or modern secular theories, but should include the teachings of the Catholic Church. As Cousin phrased it in a later work, the eclectics were

> [D]eclared partisans of all systems favourable to the saintly cause of the spirituality of the soul, of the freedom and responsibility of actions, of fundamental distinctions between good and bad, and of the selfless virtue of God, creator and organizer of the world, support and refuge of humanity.[75]

'God' as 'creator and organizer' may sound similar to a vague Deism, but Cousin went further, calling for the adoption of a republic founded on a Cartesian God who can and does interact with the material world.[76] Though not as explicit as Chateaubriand or Bonald in calling for the re-emergence of the Church, the eclectic thinkers, like many of the combatants in the war against the sciences, were linked by a strong belief in the morality of the Church, if not its literal cosmology.

In the final work of the eclectic handbook *Collections*, 'De la philosophie en Belgique', Victor Cousin (1792–1867) provided the clearest explanation of the methodology of eclecticism, refining Van de Weyer's proposal to comb the historical record for a more concise methodological approach. For Cousin, the eclectic philosopher began with establishing the *sens commun* in man. This should not be interpreted as 'common sense' in the way of clear and direct knowledge, but instead literally the sensations and feelings that are common to all people. Once the components of the *sens commun* were established through an examination of the behaviour and morality of contemporary man – an investigation Quetelet took up in *Sur l'homme* – the second step was to show how this *sens commun* originated in the mind of man. While Cousin also assumed this could be done 'scientifically', this was more of a necessary assumption than empirically provable proposition. Finally, and most importantly, the eclectic philosopher returned to the historical record of philosophy in order to research how various philosophies treated the different aspects of the *sens commun*. By focusing on what was common in the historical record of philosophies, the philosopher could then avoid the whole notion of contradictions both within and between philosophies and, voilà, avoid the criticisms of the Enlightenment.

Readings of Cousin, Reiffenberg and Van de Weyer make it difficult to accept recent assertions that eclecticism was simply the continuation of the Enlightenment dream of progress.[77] Aside from the fact that the eclectics in Belgium all explicitly mention and attack the *philosophes*, they did not share the Enlightenment belief that progress was achieved through the application of practical science. Eclecticism may have represented a dim echo of the Enlightenment call for general progress, but it was just one of several movements that failed to recapture the most crucial element of the eighteenth-century project: the pragmatic synthesis of the arts and sciences. While Cousin and his Belgian followers had criticized positivists for abandoning all other pursuits in favour of an enslaving materialism, eclectics hid behind vague notions of a scientific analysis of past philosophies. But this scientific analysis was really nothing more than a philosophical cover for the return of the Catholic Church at best and an empty spiritualism at worst. At its root eclecticism was no more a science or even a philosophy than positivism was a religion. Though Quetelet worked closely with a number of eclectics and cited Cousin often, he would have to look elsewhere for the methodology to locate the *sens commun* in the actions of mankind.

Conclusion

While on the one hand eclecticism represented a way for Quetelet to navigate through the two extremes of Romanticism and positivism, on the other hand its voidance of the scientific content of Enlightenment progressive ideals denied the two things that attracted Quetelet to the *philosophes* in the first place: a commitment to practical improvement and the success of quantitative study in the natural sciences. Furthermore, Cousin himself was moving towards the aesthetic theory of the Romantics, famously calling for 'religion for the sake of religion, morality for the sake of morality, and art for art's sake'.[78] In fact, in denying the instrumentality of art or science even to *morality*, Cousin surpassed Chateaubriand in proclaiming the independence of art. More importantly for Quetelet, however, eclecticism left no room for a *quantitative* investigation of morality except as a preliminary descriptive account that could then be used by eclectic philosophers to guide their historical study. As Quetelet would later make clear in his writings on society, statistics had the power to describe not only behaviour but also to provoke changes in the laws that governed society, a reach Cousin, Reiffenberg and Van de Weyer would not allow. While eclecticism was initially attractive to many of Quetelet's colleagues, the intellectual movement was not broad enough to contain his peculiar (for the time) mix of quantitative study and literary expression. Though its name would imply more, the philosophical creation of Cousin had by the 1840s become no more than an impoverished relative of the Romantic critique.

As eclecticism turned conservative, there remained only two cultures – represented by Cousin and Comte – that continued to be influential in Belgium.[79] Quetelet's relationship to these two was complicated, though it might make sense to rule out an affinity with the eclectic methodological programme as Quetelet's interest in ancient philosophy, sixteenth-century rationalism and Catholic cosmology was close to non-existent. And while Quetelet and Comte were certainly not similar in their methodologies, and Quetelet's position as a positivist is far from certain, it is indicative of the complications that arise in understanding Quetelet as a descendent of the *philosophe* social theorists that many of his close friends and influences rejected the Enlightenment programme.[80] The criticisms of Enlightenment made by Quetelet's close colleague Reiffenberg cannot be simply dismissed as two friends who had different intellectual approaches – Quetelet's later frequent quotations of Cousin would indicate at least some appreciation of the baron's philosophy. To complicate matters further, the intellectual tradition with which Quetelet at times seemed to identify – German-inspired French Romanticism – had few institutional or academic homes in Belgium and certainly none for a student whose dissertation was on a mathematical theory of the graphic representation of the interaction between flat surfaces and cones. As the war was winding down in continental

Europe between the sciences and positivism on the one side and the arts and eclectic spiritualism on the other, Quetelet was in the odd position of being trained mostly in the tradition of the former but inspired by the latter. Quetelet hinted at this dilemma in his poetry but never directly commented on just how contradictory and confused his world was at the beginning of the 1820s.

The conflict of cultures would in fact never be resolved by an internal reckoning on Quetelet's part. Instead it was the practical needs of Quetelet's state patron during the next two decades that helped him sort out his various influences.[81] As seen in the next chapter, Quetelet was able to produce a significant body of literary work in spite of the demands of state employment. Unfortunately for Quetelet, Cousin could not end the war of sciences and letters, and eclecticism had far less value to the government of the United Kingdom of the Netherlands than a thorough and accurate statistical calculation of its population and resources. Lacking the meta-historical judgement to untangle this web of intellectual influences and contexts, Quetelet instead allowed the practical demands of the state to cohere his ideas for him.

2 CASUALTIES OF WAR: QUETELET AND FRIENDS IN GHENT AND BRUSSELS, 1815–23

The Romantic Years

Was Bonald right? Had the war between the arts and sciences really made things that difficult for artists and men of science? Did the rhetoric of Chateaubriand, or the institutionalization of science in Napoleonic France, overstate the degree to which scientific and literary ideas were kept separate? Was it still possible to pursue a wide variety of interests and succeed in creating new scientific ideas? In many cases throughout Europe, the answer to this final question at least was an unequivocal yes. Amateur scientists abounded throughout Europe and continued to make important contributions throughout the nineteenth century. Even in fields like astronomy that were becoming more mathematical, more standardized and more expensive, unaffiliated and unpaid men and women of science made their mark. In one recent account, the amateur astronomer William Huggins is noted for his success even as he pursued the 'exotic rather than the mundane'. As Barbara J. Becker noted, Huggins succeeded in large part because 'As an independent observer he was free of the obligations and commitments that restricted his institution-bound contemporaries'.[1] Huggins was not alone, and there is little question that men of science like John Herschel and Alexander von Humboldt continued to pursue science outside of the 'obligations and commitments' to institutions, oblivious to any war of the arts and sciences.

As robust as amateur science remained throughout some European countries, the exceptionalism of Huggins, Herschel, Humboldt and others points to the larger predicament of 'institution-bound contemporaries' in the sciences, particularly those in countries with a less developed sense of independent amateur science. In particular the newly created United Kingdom of the Netherlands in 1815 was firmly committed to reforming the Catholic lowlands through the creation of new schools and government-sponsored institutions. As will be seen in Chapter Three, Quetelet was at the forefront of this movement to institutionalize science and scientific education in Belgium, yet the move to embrace a Napoleonic vision of state-supported science was not a pre-ordained outcome for the young Belgian. As his early interest in Chateaubriand and de Staël show, Quetelet had at least some sympathy for the romantic vision of 'art for art's sake', and he spent much of his teens and early twenties composing verse and reflecting on the plight of what he called the 'troubled author'. Furthermore, Quetelet's network of friends in his birthplace of Ghent and adopted home of Brussels tended towards the bohemian, as he lived in a series of houses frequented by authors and painters, including Jacques-Louis David. While Quetelet might seem to have naturally moved from a dissertation in maths to creating statistical and observational institutions of science, the period of Quetelet's 'romantic years' and the history of his earliest network of friends complicates this image.

The two biographies written on Quetelet close to a century ago would not support such a focus on the Belgian's early years; his reflections on Romanticism, his literary output and his circle of friends in Ghent receive only brief mention, and then only as a biographical introduction.[2] Yet the period between 1815 and 1823 cannot be neglected as a mere pre-history of a statistician. Instead, an investigation into Quetelet's interest in Romanticism and his own literary writings can lend insight not only into the direction of his later statistical theory of society, but also serve as an example of the effect Bonald's 'war' had on men of science and men of letters. It is of course unknown whether Quetelet would have chosen to remain both a mathematician and writer like d'Alembert, but what is known is that the intellectual context of the United Kingdom of the Netherlands in the early nineteenth century introduced new practical and professional consequences for making such a choice.

Quetelet's operas, poetry and essays may not have signalled the arrival of a great writer, but they were significant in his time and place. A summary and selection of this work reveals that Quetelet was both a serious student of the arts and may have been working out some of the conflicts between being an artist and a *savant* in Belgium at the time. It also suggests that he was not moved towards the sciences because of interest alone. In comparison to Quetelet's relatively seamless move into the state-sponsored sciences of the United Kingdom of the Netherlands, three of Quetelet's closest friends from Ghent and Brussels – Germinal Dandelin, Louis-Vincent Raoul and Reiffenberg – struggled to adapt to the reforms in the Low Countries. Their collective experiences confirm that Bonald's war had real consequences for practising science in Belgium.

The Troubled Author

From the ages of nineteen to twenty-seven, Quetelet composed at least forty poems, sixteen opera librettos, countless translations of classical poetry, a eulogy to the painter Odevaere and one essay on the history of the Romantic genre. As a whole, it is a decidedly uneven output ranging from hackneyed efforts at imitating sixteenth-century bards to deeply moving verse on the death of loved ones.

Yet most of the poetry found its way into local journals, the *Essai sur la romance* was widely distributed, and one opera libretto Quetelet co-wrote was produced with an accompanying score by Charles Ots, a notable Belgian composer. Though he was far from famous, the entirety of his work cannot be dismissed as a hobby and there was no indication from this time that he privileged his scientific work over his dramatic works, or even that he saw a division at all. If anything, to judge by a later admirer, Quetelet in his early twenties 'was much more preoccupied with poetry than with science'.[3] Furthermore, his eulogist Mailly reported that after being named professor of mathematics in conjunction with acceptance to the University of Ghent, Quetelet was happiest because his new position as a student professor allowed him the freedom 'to occupy himself with art, science and literature; to design, to play the flute, to read Pascal, to study Newton, and to compose verse'.[4] Though a more recent claim that Quetelet's entire body of work can be seen as 'essentially poetic' may push the point too far, the quantity and quality of the work at this time are indicative of a serious student of the arts.[5]

As a writer, Quetelet's primary strength was his broad knowledge of languages. In January of 1815, three months before the notification of his appointment at Ghent, Quetelet was asked to join the local chapter of La Société anglais, a group who defined themselves as 'an association of men who love and cultivate the arts and sciences' and who desired 'a perfection of a language already well utilized'.[6] One requirement was fluency in English, a criterion Quetelet had no trouble meeting, as it was just one of the seven modern languages he could read in addition to Greek and Latin.[7] In fact, to go by his private notebooks from the time, Quetelet seemed to do little else but translate, including pages of Byron, Schiller, Ovid and Horace. He was welcomed enthusiastically into the group.[8]

While Quetelet eventually published several elegiac poems in the style of Horace, his first artistic break was as a dramatist. Written in collaboration with the student who had beaten Quetelet in the second grade maths exams, Germinal Dandelin, the opera *Jean Second* premiered in Quetelet's hometown on 18 December 1816 at the recently built Théâtre de Gand. While both the Quetelet and Dandelin families were well connected in the city, the decision to stage an opera written by two students in their early twenties was apparently driven by merit alone. Ots had agreed early on to score the work, and it was this independent decision, rather than any outside influence, which secured a spot on the programme.[9]

Set during the reign of the Holy Roman Emperor Charles V, *Jean Second* centres on Alfred de Rosenberg, a Belgian aide to the emperor who has been framed for murdering his best friend and has fled to Charles's Belgian château in Ghent in order to beg for mercy. Though Rosenberg believes that he has actually killed his friend, the murder had been organized by jealous rivals seeking to improve their own position in the court. On the same day Rosenberg arrives in Ghent, the friend's brother, Alvarez, has come to find his sibling's assassin. In the opening scene, Rosenberg describes the confusing events of the 'assassination' in detail to Emma, a friend employed at Charles's estate. The dead man, though 'noble and brave', is described as having a poor temper, and Quetelet makes explicit the differences between Belgian and Spanish temperaments. Rosenberg explains the ensuing drama to Emma:

> It was not difficult for us to quarrel. His insolent arrogance outraged me ... I am Belgian, Emma; more proud perhaps because my country was insulted and I would avenge it. The blood of the Spaniard ran on my sword ... They accused me in the presence of the King and said they saw me covered in the coat of an assassin, cloaked in the shadows of the night and plunging a poisoned dagger into the breast of the unfortunate Lémos ... My head was promised to his family. I looked in vain to justify myself but my voice could not make it to the king. I had to escape; I had to leave to avoid an infamous death. Alone, without friends, without assistance, lost in the forests, I left Spain immediately to return to my country and arrived here in Ghent.[10]

Emma warns Rosenberg that Alvarez has already arrived to avenge his brother, but believes that the poet in residence, Jean Second, can help plead his case. As the intrigue develops within the house, the reader learns that while Rosenberg is in love with Emma, Emma is in love with Jean Second; the only reason she has refused to admit it is because she believes Rosenberg's love is the only thing that sustains him. The situation dawns slowly on Rosenberg:

> Do you think I ignore that Jean Second, the famous poet, has professed his love to you; it despairs me. I learned of it before I arrived in Ghent and since then I have had no rest: jealousy devours my blood; twenty times during this pain I have wanted to kill myself; but I have hoped always to only be tormented by a terrible dream.[11]

If nothing else, *Jean Second* shows the work of a writer steeped in the conventions of the Romantic genre: unrequited love, injustice and hopelessness. All of the characters of the story save Charles suffer some tragic loss, and the scenes of violence and heartache are full of vivid imagery: a 'plunging dagger', an assassin 'cloaked in shadows' and a protagonist 'tormented' by suicidal nightmares. Yet unlike his literary heroes at the time, Goethe and Schilling, Quetelet did not bring the production to a tragic end. Instead, Rosenberg is saved through the poet's clever speech to the emperor, Alvarez spares the framed Rosenberg and is rewarded by Charles with the governorship of Madrid, Jean Second is added to Charles's circle of advisors and Rosenberg and Emma live happily ever after. Demonstrating the unceasing optimism that would persist in his social theory, *Jean Second* seems an early example of Quetelet's blend of Romantic form and Enlightenment content.

Striking a similar balance between loss and redemption – or the tragedy of the past and hope for the future – is *Moschur*, an unperformed libretto Quetelet wrote between 1816 and 1817 about a woman who, because of a vicious storm, mistakenly believes her fiancé has abandoned her.[12] Though Quetelet dealt with loss in almost all of his eleven completed librettos, *Moschur* was his first work in which the forces of nature act as the primary metaphor. Given his later work in meteorology, it is not hard to envision this work as a balance between the unforgiving nature described by the Romantics and the possible harnessing and taming of nature found in the most ambitious of the new meteorologists.[13]

The first act of *Moschur* finds a 'dismayed' Lilière on the shores near her home, lamenting to her father Moschur that her fiancé has been killed in the storm: 'All your efforts are hopeless / I have lost Cléomène / I fear death little / All reason is in vain / do not try to stop me'. Moschur, admitting to his weakness in the face of nature responds: 'I succumb to this inhuman storm / it has triumphed over my best efforts'. Near suicide, Lilière is finally convinced by her father to return home because her lover will never come. Cléomène does, however, make it to shore, but only to see that no one is waiting for him. Cursing 'unhappy love' and convinced 'she has forgotten [him]', he contemplates killing himself. Here again, however, Quetelet refused to submit completely to the power of nature ('all reason' may *not* be 'in vain'). Near the moment when both lovers are close to suicide, Moschur observes a break in the storm and sets out again to find his daughter's lover without her knowledge. Finding Cléomène, Moschur rejoices and quickly takes him back to find Lilière, but when they arrive they find her lying motionless and for a moment the reader believes she is dead. Yet just as Cléomène is about to despair, his heavy-eyed lover comes to life. She had only been napping.[14]

Throughout the early operas, Quetelet returned to the metaphor of the storm as the chief obstacle to the artist, but in a final libretto, written in 1819, he demonstrated scepticism towards the artistic pursuit in general. *M. Dièse ou l'auteur dans l'embarras*, as the title indicates, highlights the problems of the artistic life, a possible sign that Quetelet was concerned about his own ability to continue to write literature in the political climate of the United Kingdom of the Netherlands. The plot concerns the bumbling efforts of a young poet to secure the love of a woman, Julie, in opposition to the wishes of her sister, Lisette. 'A word only', the poet Dorante asks repeatedly, only to hear Lisette's curt responses: 'No ... she ignores you, she detests you, she hates you'. To try to win over Julie and Lisette, Dorante enlists the help of his friend and an older composer. Dièse, the 'troubled author' of the title, declares that 'love is a war' and that Dorante must stop at nothing in pursuit of his love. Dièse is a gifted poet, and in a scene reminiscent of Cyrano de Bergerac, agrees to provide Dorante with the words to win Julie's affection.[15]

The frustrated artistic ambitions of Dièse seem to be Quetelet's main concern, and it is not difficult to discern in Dorante and Dièse Quetelet's potential vision of his present and future. The fictional Dièse, like Quetelet, had composed operas for several years but had only managed a modicum of success. By 1819, the year in which *M. Dièse* was likely written, Quetelet had finished eleven operas and started another five but had yet to see another work performed after *Jean Second*. Though Quetelet later joked about his early operas, the language of *Dièse* may offer a better picture of his mind-set at the time.[16] Lisette in fact offers a bitter portrait of the life of the artist in warning her sister about a potential artist as a spouse. As a professor at a small university with little to no income from his writing, Quetelet was not far from the 'poorhouse' described in the following passage:

> If He could make us in His head, each would be last a musician [or] a poet. [But] thanks to the heavens, [Dorante] does not have a poetic mind, like a worker in comedic verse, always rhyming, rhyming, rhyming... too many of these songs will take you at last to the poorhouse.[17]

Lisette therefore believes Dorante might be an acceptable match for her sister, if only because he lacks the talent to truly find a career in the arts and will eventually be forced into more lucrative practical employment. The tension between the need to work and the drive to create art are found in the many conversations between Dièse and Dorante. The latter is interested in the benefits of the arts, but admits that Dièse is the 'superior mind' whose 'modesty hides a supreme genius'. Dorante realizes that he does not have the old man's dedication and instead only wants the attendant fame of the poetic life without the sacrifices. Dorante echoes the claims of Chateaubriand that the work of an artist will be 'fixed in posterity' and that Dièse 'will find a place in the temple of memory'. Unwilling to make the practical concessions to the life of the artist, however, Dorante is content to use Dièse's words for his own objectives and 'to associate' himself with the 'honour' of Dièse's 'glorious name'.[18] He could avoid the serious work involved in poetry and still achieve fame through a connection with the poet. Between Dièse and Dorante, Quetelet captured both the Romantic view of the sacrificial artist as Chateaubriand imagined and the practical considerations of the new professional man of science in the nineteenth century. Though the myth of the struggling artist was only a recent creation, it was a myth that both sides were willing to believe.

In the final act of *M. Dièse*, however, Quetelet questioned the received Romantic view of the artist, indicating that his allegiances to the sacrificial art of Chateaubriand may have been on the wane. Dorante, frustrated with his own work and concerned about failing to win over the girl, pleads with Dièse to read the young man's recent composition. Dièse balks at reading the entire work, replying that Dorante should 'just give me the name' of the opera.

> DORANTE. The title only?
> DIÈSE. Yes.
> DORANTE. It is ... The Author ... The Author ... in ... Trouble.
> DIÈSE. I pity this author.[19]

As Dorante presses on with the description of his work, it becomes clear that the old age and struggles of the 'author' are not his own, but those of Dièse. Dorante, in attempting to explain his own struggles with the artistic life, is actually commenting on the failures of the older man. The 'troubled author' then is not the dilettante who refuses to give up the life of the family for his artistic pursuit (Dorante), but the Romantic who pursues immortal fame against all odds (Dièse). In a clever move, Quetelet inverted the expected relationship between the wise older poet and the practical younger man that he had created in the opening scenes. It is Dièse, the committed poet, who is troubled and who is to be pitied, not the uncommitted and pragmatic Dorante. Given that the title of Dorante's proposed opera is the same as the subtitle to *M. Dièse*, it is not hard to believe that Quetelet shared Dorante's vision of the poet's life. As a mathematics professor who also had hopes of publishing his artistic compositions and essays, Quetelet may have realized that this kind of balanced artistic life was no longer possible and that a life in Chateaubriand and Lamartine's wilderness was not for him.

M. Dièse was Quetelet's final opera. Interestingly, it was also the only opera in Quetelet's archives in the Académie royale written in a hardbound notebook. While all of his other remaining works consisted of a constant series of alterations in soft notebooks and scrap paper, this final dramatic piece was written in a neat script in a sturdy volume. Did Quetelet believe it to be a more significant piece than his earlier works? Given his lack of later reflection on this period of his life, it is hard to tell anything of his plans, yet the new medium for his compositions is suggestive. If Quetelet did see his work as maturing to the point of serious art, then the practical conditions of its composition stand in a strange dialogue with the thematic elements of the work. Seeming to give up dreams of artistic immortality in the content of the opera, the physical form might suggest that Quetelet held out hope. As the final work in an intense period of creation, *M. Dièse* revealed the young writer's attempt to balance a professional career with the ambitions of an author.

It might be wrong to read too much into the twenty-three-year-old Quetelet's thoughts from the three librettos of *Jean Second*, *Moschur* and *M. Dièse*, but it seems clear he was engaged in debating the benefits and consequences of an artistic life. The rapid development of thematic complexity from *Jean Second* to *M. Dièse* – from simple mimicry of the tropes of the romantic bards to a self-referential drama on artistic strife – shows a remarkable progress for a writer so young. Quetelet gave no reason why he stopped writing operas, but it is unlikely it was because he simply lost interest in the form. Perhaps, like Dorante, Quetelet may have been thinking about the hardships an artistic life would entail when combined with the demands of his university work. Though marriage would still be another seven years away, Quetelet was aware of the practical considerations that so troubled his character Dorante. Even if it was not a conscious decision to abandon writing for a career in government science, the pressures of his professional career were certainly encroaching. Because his appointment at Ghent was as a student professor, 1819 was not only the year of his final libretto; it was also the year of his dissertation.[20]

Poetry and Reason

While Quetelet's operas displayed a deep immersion in Romantic themes, and his dissertation a foreshadowing of his future career, his poetry from 1817 to 1822 is more difficult to classify. Of note is that he continued to write and publish poems even after leaving Ghent to join the faculty at the Athénée de Bruxelles in 1819. While possible that the demands of poetry were not as great as dramatic works – which ceased once he left Ghent – some of the longer poems fill several notebooks. A more plausible explanation is that Quetelet found his voice more in the evocative works on nature than in the narrative demands of opera. There was also very little of the optimistic writer of *Jean Second* and *Moschur* in these later poems. Instead, the themes became darker and the contrast more apparent between optimistic pronouncements on the possibilities of a perfectible science and scepticism of the limits of human nature. Quetelet said little explicitly about this tension in his own writing, but it does appear that Quetelet abandoned his poetry at the time when it was at its most despondent and, importantly, most at odds with the defining aesthetic of Romanticism. The poetry written during these years not only challenges the optimism of his later works on the power of quantification, it also suggests why Quetelet could not follow his idol Madame de Staël into the avant-garde of German-influenced literature. Quetelet's was a poetry that had an affinity with neither the optimistic pronouncements of utopian positivists nor the formal condemnation of science of the Romantics.

Quetelet certainly had more publishing success in poetry than he did in writing librettos. In all, there are at least thirty-one completed poems, though records indicate the total may have exceeded forty.[21] They range in date from 1818 until the 1840s, when he composed several *éloges* for his friends and family, including moving verses dedicated to the passing of his sister and nine-year-old daughter. While the early poems mark a continuation of the themes of loss, abandonment and alienation from nature, many of the works for the first time contain references to science and scientific luminaries, with the praise once reserved for Homer, Virgil and Horace equally distributed to Pascal and Newton. The poetry discussed here, while still formal and conventional, shows evidence that Quetelet's emergent career did not yet inhibit an interest in the fine arts. On the contrary, it had initially encouraged it.

Quetelet's publishing breakthrough came through the aid of his friend Louis-Vincent Raoul. In 1818, Quetelet's 'La Veillée des bardes' appeared in Raoul's *Annales Belgique*, a small journal that in spite of its ambitious title mostly published material from Ghent and the surrounding area.[22] Quetelet followed with a poem entitled 'La Poète mourant à son lampe', an accomplished piece that continued with the tragic themes he had developed in the later operas. The poem recounts the pleas of a poet to his 'guiding flame' as he confronts a premature death.

> The poet sang, and his sweet face
> Leaned towards the light, fallen and pale
> Hating the slowness of pain which consumed him
> His plaintive voice expired in a moment
> And he rose again and with his weakened sight
> Looked in a dying moment towards his dear lamp
> The lamp was extinguished ... it was already out.[23]

Though formally the work was conservative, the tone marked a departure from the common poetry of the time and would anticipate some of Quetelet's most persistent themes: death, suffering and the limits of investigation. The lamp, of course, is also a symbol of knowledge. At the same time the lamp's flame is dimming, so too are the poet's hopes of receiving an answer. Just twenty-two at the time the poem was written, Quetelet displayed little of the optimism of finding knowledge in these early works that would later become crucial to his social theory.

From the perspective of his later writings, Quetelet's poetry also unexpectedly examined nature as a mysterious and elusive entity. Like the troubadours he admired, Quetelet's poems as a whole constituted a search less for a true understanding of nature than as a way to cope with the dangers and challenges nature presented. Like his later work in meteorology, geology and *physique sociale*, understanding nature meant not necessarily the discovery of iron laws, but rather understanding enough of the dynamics of a system so that some effects could be altered. In his poetry on nature, Quetelet returned consistently to the metaphors of light and darkness, storms, foliation and the changing of the seasons. Though he eventually tried to understand these forces through the collection of data from a vast international network of scientists, at this point he was content to examine them in verse.

Like the storm that almost cost Moschur the life of his daughter, weather often appeared in Quetelet's poetry as a frightful force that prohibits knowledge. In 'La Comtesse Ide', a man searches for his lost lover but loses track due to '[t]he snow that covered the earth / and like a funeral shroud / turned the side of the valleys white'. It is only with the thaw of spring that he could return to his search and pick up the forgotten clues. Less fortunate is the lover Edwin from 'Le Scalde et Lysis' who fights the heavens in his 'rain battered' craft but never returns home. Instead: 'In vain his arms struck the waves in fury / The wave engrossed the entirety of his heart'. Quetelet's poetry found mankind often at the mercy of nature in pursuit of knowledge and progress. Only when the storm has ceased, or the season turned, can work be resumed. As Quetelet wrote in 'Épître à mon sœur', knowledge can only appear like 'the rays of the fertile heat' that 'come to rejuvenate the world after a long winter'. In these poems, all likely composed between 1819 and 1822, Quetelet extended his theme of a merely indifferent and frightening nature to one that was unconquerable. The happy endings and fortuitous breaks of the early operas, when both mistakes and storms cleared up, were replaced by an inscrutable nature.[24]

The most revealing examination of the competing ideas of light and darkness occurs in an unpublished and perhaps unfinished poem entitled 'L'Illusion'. The protagonist, a hermit wandering in the forest 'embraced by an air of gloom', is one of Quetelet's most powerful characters. Lost in the woods, the man overhears a funeral procession in the distance; the sounds of weeping women are transformed into 'rumblings of storms' that 'haunted before him in the heavenly blue'. The forest and storm begin to swirl around him and the hermit becomes aware of a transformation in his surroundings, as if he had left mortal earth: 'I crossed the limits of the world / Lifted without effort to the realm of the elect'. Believing he has died, the man anticipates bodily transformation from the darkness and gloom of the earth to the knowledge of 'light of the heavens'.

> I was going therefore to be intoxicated by the other life
> To be transported to the celestial concert
> I was going to see the highest of infinite glory
> And know the infinite laws of the universe.[25]

But it is not to be. Before passing into the new world, a phantom appears before the hermit and obscures the light with mist. He tells the man that he is not yet dead and that he will have to return to earth because the living world is better than he realizes. Before the man can question the phantom, 'the mist which proceeds daybreak and makes all before the day, had disappeared'. Upon returning to earth, he joins the funeral procession and cries along with the women. At last he gains the courage to peek inside the casket, relieved to discover that the body inside is not his own.

This 'magnificent deceit' of rain and clouds provided by the phantom keeps the man from discovering the 'infinite glory' of universal knowledge but, conversely, allows him to appreciate more his own mortal existence. Quetelet had certainly been exposed to the Enlightenment effort to discover and learn about the 'infinite laws of the universe', yet 'L'Illusion' presented a deep scepticism about the efforts to find these laws on earth. The general programme which Quetelet would pursue in the next few decades – discovering the laws of society through statistical description – may seem contradictory to this scepticism, but it would be well to remember the lessons of the phantom as Quetelet defended himself from charges of materialism and determinism for his later work. Man's fantasy of a universal law was not only a possible illusion, but the claims of true knowledge on earth could be read as a deceit as well. Stripped by the phantom of illusions of universal knowledge, the man can then return to the world with an appreciation of what he *can* understand.

While Quetelet eventually comforted himself with knowledge from his statistical surveys and astronomical observations, in his twenties art was as likely as science to assuage his fears of the unknown. In 'À mon ami de Reiffenberg', composed in the early 1820s, Quetelet confessed that the 'muses' of the arts 'charm[ed] my spirit' but also 'clear[ed] up my reason'. They would give comfort to a future in which 'death awaits'. Quetelet's poem continued as a broad homage to artistic inspirations, containing possibly his most powerful statement on the value of art. 'On a satisfying journey, I march without effort', he began:

> The love of the arts, by the their immortal flame
> The pleasures more sweet, the pain less cruel
> I have known unhappiness, and I give thanks to them
> Unhappiness seeks to flee at first
> From my numerous friends
> It stays alone ... for me, I prove the contrary
> I have come to believe that which I see in the arts
> Consolatory friends
> I offer my regards to their diverse treasures and wonderful marvels.[26]

Quetelet wrote often of the 'consolatory' effects of poetry and literature, and almost all his later *éloges* would include some original verse or, at the least, an epigram from a classic poet. So too were most of his early friendships marked by an appreciation of literature. But this 'love of the fine arts' would eventually recede as Quetelet turned to astronomical and statistical investigations to relieve the anxieties produced by the uncertainties of nature. The 'numerous friends' would no longer be the poets and artists with whom he lived in Ghent, but the hundreds of men of science and administrators he would later correspond with.

Quetelet specifically mentioned the relationship between science and art in a poem dedicated to another friend, his co-librettist on *Jean Second*, Germinal Dandelin. In a piece written in 1823, around the same time he travelled to Paris to meet the leading French mathematicians and astronomers, Quetelet praised Dandelin, who 'in an instant could abandon all this abstract calculus' and 'place the same balance between the side of pleasure and the other of suffering'.[27] Yet notably, Quetelet also lauded the work of Leibniz, Gauss and Laplace for uncovering the 'laws of the universe', a possible departure from the 'L'illusion' written four years earlier. Telling Dandelin that he loved most their school days when every day brought a 'new pleasure in the grand fields harvested by Euclid', Quetelet for the first time spoke of the great mathematicians in the same manner as the great poets: 'I have taken their virtues and noble exploits to my heart'. The 'Épître à Dandelin' then might indicate a juncture between the practical applications Quetelet would eventually create from Gauss and Laplace and the 'pleasurable' inspiration he took from them as a student. He even spoke of mathematical work in 'understanding the laws of the universe', the same laws that in 'L'Illusion' seemed to be reserved for the heavens. Four years after leaving his hometown to take a position as professor of mathematics, Quetelet also seemed nostalgic for the Romantic innocence of his youth, concluding with the line that he will always 'carry my childhood reverence for *la Chevalerie*'.[28]

Was Quetelet hinting at his new responsibilities by locating his 'reverence for *la Chevalerie*' in the past, even as he promised to hold onto it? It is possible that he was beginning to see the opportunities available in the sciences in the United Kingdom of the Netherlands as he confronted his own struggles as a poet. In 'Épître à M. Tollens', he complained that the contemporary Gantois poet 'hid under a vast shroud' that overshadowed other poets from the city. Quetelet confessed that he was both 'inspired' and 'chained' by Tollens's words. Even though Quetelet's work was impressive, the overriding voice of his poem recalls a student who cannot receive the answers he wants from a professor.[29] After telling Tollens how he has sought to model his own work after him, Quetelet concluded: 'I am therefore condemned; with you receiving more praise I will be less capable.'[30] Quetelet had written of his own struggles before, yet the construction of the final clause in the future tense indicated the young poet's resignation.

If Quetelet's internal debates between the worlds of science and art were still at the level of the subconscious in his poem on Tollens, Quetelet made the contrast explicit in an 1821 composition entitled 'La Poète et la Raison', a work that stands in interesting relation to not only Bonald's comments on the war of art and science but also to the dichotomies drawn by Pascal, d'Alembert, de Staël and Chateaubriand. Inspired by reading a book on the great poets of the medieval Netherlands, Quetelet pitted an imaginary Poet and Reason against one another in a discussion over the proper place of self-sacrifice and reason in the production of art. The piece begins with Poet concerned about a new work of verse he feels in his heart but cannot express. Overhearing the lament, the character Reason instructs him that art can only come about with his help and that, in any case, Poet should not worry; it will be up to the scholars to decide who is great. Poet objects, however, listing the great names of Belgian verse, but at each point Reason responds: 'but who engraved their great name on the temple of memory'. Poet, becoming more enraged, scolds Reason for his 'vulgar' diplomas but in the end is resigned to his fate. Like d'Alembert in his 'Essai sur la société des gens de lettres de des grands', Quetelet concluded in one of his final literary pieces that art must always be dependent on reason.[31]

Though Quetelet published occasional poetry through the Société de littérature in Brussels, 'La Poète et la Raison' seems a good place to finish this survey of his artistic production. In this short burst of five years of creative activity in dramatic work and poems, Quetelet displayed a striking progression. While the early operas are mostly retellings of classic troubadour stories of mistaken identities and unrequited love, his later works show an appreciation for the new ideas forming around Chateaubriand and in the literary circles of Paris. Rather than creative exhaustion or stasis, 'La Poète et la Raison' finds Quetelet asking some of the same questions posed by famous writers and artists at the time: could art be rational? Could it be purposeful? Or were the poets doomed, like M. Dièse, to pointless obscurity and misery? Quetelet's debate between Poet and Reason indicated that art could not have purpose by itself, and though the French and German Romantics gladly took up the mantle of 'uselessness' in opposing themselves to the careerism they found in the sciences, Quetelet was not willing to go that far. He could not become the 'troubled author' he foresaw. All of his later writings in astronomy, statistics and demography, in fact, could be described as universally *purposeful*. Quetelet's shift towards purposefulness may have been fine for writers of the eighteenth century like Diderot or Voltaire, and it helped make his career in the Belgian state, but it would not have found favour at the editorial offices of Chateaubriand's *Mercure de France*. Though Quetelet shared many of the same influences as the Romantics he copied and admired, a review of his operas and poetry reveals that he was not willing to follow the road of Chateaubriand and Lamartine into the extremes of 'art for art's sake'.

A House in Ghent and Brussels: Dandelin, Raoul and Reiffenberg

Part of Quetelet's ambivalence towards his own artistic development may have come from his friends' struggles once he moved from Ghent to Brussels in 1819 to teach at the Athénée de Bruxelles. As the poems on Reiffenberg and Dandelin suggest, Quetelet's interest in poetry was tied intimately to his relationships with his friends. Not only was his friend Raoul the most likely to publish his works, but the diversity of opinions in this group expanded his interests in a number of ways. Though his closest friends Dandelin, Raoul and Reiffenberg did not enjoy Quetelet's success or international renown, each was a highly connected and influential figure in local literary societies, scientific organizations or both. Less famous certainly than the French *savants* and German observatory directors Quetelet later met, the impact of the Ghent and Brussels networks were as least as relevant on the course of his career as his later encounters with the heavyweights of the new professional scientific elite.

As the only student to best Quetelet in their childhood maths exams, Germinal Pierre Dandelin (1794–1847) seemed destined to take up a similar high position in the budding French and Belgian bureaucracies of the early nineteenth century. Born two years prior to Quetelet in 1794 in Le Bouget outside of Paris, Dandelin moved to Ghent in 1807 to attend the *lycée* and immediately excelled in the sciences. Quetelet describes the two at this time as 'inseparable', pursuing overlapping studies in 'sciences, literature, or *beaux-arts*'. Dandelin not only was one of the best design and math students in school, but he also had 'a strong feeling for art' and played an excellent violin.[32] But the 'necessarily incomplete education' that Quetelet had complained about forced a decision on Dandelin when he was eighteen. He needed to specialize in a single career and in 1814 chose to enter polytechnic school to pursue a degree in mechanics.

The extra years the Frenchman had on Quetelet would prove important, however, as the period was not a favourable one to begin professional study in Belgium if old enough to serve in the military. At nineteen, Dandelin joined with the National Guard and attained the rank of sergeant in defending the island of Walcheren in 1814 against a British invasion. A year later, with Dandelin about to return to school, the return of the exiled Napoleon delayed his career again. After earning the medal of the Légion d'honneur for his fighting during the Hundred Days, Dandelin briefly considered a career as a military officer in the new army of the Netherlands.[33] Instead, in a move that would anticipate much of his peripatetic career, he returned to Ghent where he and Quetelet began their brief partnership writing operas.

Dandelin, perhaps even more than Quetelet, seemed destined for a career in the arts. Quetelet claimed of his friend that he 'spoke with charm and an infinity of visions', and Quetelet's poem to his friend is full of reminiscences on their shared love of art.[34] Their co-written *Jean Second* had been successful enough to attract the attention and support of Ots, and the performance appears to have received decent enough reviews during its first two nights.[35] Dandelin was a regular contributor to the artistic groups in Ghent, and there seemed to be no limit to the man who dreamed 'of things supernatural, and created numerous theories with his brilliant imagination'.[36]

Yet Dandelin did not enjoy a successful literary career, and to judge by his later letters to Quetelet, was frustrated by the professional ethos of the day.[37] Instead of continuing to write with Quetelet, Dandelin embarked on a two-decade career that was filled with government-appointed positions in the industrial sciences, work which, at least according to Quetelet, left his friend often unhappy and unfulfilled. On a mining mission to the town of Venlo on the outskirts of the kingdom, Dandelin complained to his friend that there were only 'mediocre writers' on his 'rotten journey' and that: 'I lack the means to maintain my character'. Dandelin saw in his professional work an 'atmosphere' that 'deprive[d]' him of his 'intelligence'[38] and decried the 'selfishness and greed' of his co-workers.[39] Towards the end of his career, Dandelin became so distraught that he burned almost all of his writings, most of which had been scattered and unfinished. In one of his last notes to Quetelet before his death in 1847 at age fifty-three, he confessed that he 'had forgotten the subjects of humour and melancholy' the two shared as teenagers.[40]

How had the 'brilliant' mind of Dandelin fallen into such despair? Perhaps the fault can be placed with Dandelin's father, who threatened to bring his friends to disrupt a performance of the opera *Jean Second* in order to steer his son away from an interest in the arts.[41] Yet Dandelin had sufficient talent and ability in the sciences that he should have been happy in any career choice. The comments to Quetelet suggest that perhaps Dandelin may have been suited for an earlier age of scientific exploration: as Dandelin wrote in his dissertation, 'imagination' was necessary to good work in the sciences, provided that a 'steady calm' was there to 'direc[t] the imagination'.[42] In the United Kingdom of the Netherlands, however, imagination was far from the prime motivation for William I and his advisers. As seen in Chapter Three, the jobs and careers available in the sciences were largely directed towards emulating the French system of technical and practical knowledge. While Quetelet, it turned out, was perfectly suited for this role, one of his oldest friends could only lament the lack of options for a librettist turned mining surveyor.

While Dandelin's career was troubled by an inability to reconcile the competing drives of an artistic spirit and a career in the industrial sciences, his fellow Frenchman Louis-Vincent Raoul (1770–1848) remained an unreconstructed *homme de lettres*. Raoul, twenty-four years Quetelet's senior, was the leading publisher of poetry in Ghent and co-founder of the *Mercure belge* – modelled after Chateaubriand's French Romantic journal *Mercure de France* – and the *Annales Belgique*, two journals which accounted for the majority of Quetelet's publishing career.[43] Born in Poincy, forty miles east of Paris, Raoul spent his early years alternating between teaching posts and the army, becoming a professor in nearby Meaux by the time of the French Revolution. Drafted into the army during the series of wars that followed, Raoul eventually became treasurer for his regiment. After winning some modest renown for a eulogy to the French General Lazare Hoche as well as the successful production of his play *La Chute de Robespierre*, he became director of the library at Meaux in 1807, a position he held for almost a decade.

With a distinguished military and literary background, Raoul may have been content to live out his years teaching in the French countryside were it not for a rare opportunity to the north. In 1814, after the end of French occupation, the new United Kingdom of the Netherlands had founded three new universities in Ghent, Louvain and Liège to reinvigorate (or, as some saw it, de-Catholicize) the educational system in the southern provinces. As Quetelet's appointment as student professor to Ghent at the young age of nineteen indicated, the ambitions of the royal programme were greater than the native talent on hand. Therefore, in a move Quetelet dryly described as 'not very consoling to our national pride', King William advertised throughout continental Europe to fill the vacancies, attracting not only Raoul from France, but also the eclectic philosopher Baron de Reiffenberg from Germany.[44]

As fellow members of the first faculty at Ghent, Raoul and Quetelet could not have been more opposite in background and philosophy. The former was forty-four, an experienced military officer, monarchical sympathizer, classicist and professor of literature; the latter was nineteen, naïve, a republican, modern Romantic and professor of mathematics. Though Quetelet at the time was still a student infatuated with the modern writers Schiller and Goethe, Raoul was often critical of German Romanticism, dismissing Quetelet's promotion of the Germans with the pithy line: 'the classics are those which have made the grade, the moderns are those which have not'.[45] Raoul also showed only cursory interest in the sciences, supporting his friend's efforts but never going into constructive criticism like Dandelin or Reiffenberg. Raoul's later letters are unique in this respect: most of Quetelet's correspondence in the 1820s was filled with diagrams and equations while Raoul's are bursting with exclamations for the work of Molière and with lines spilling over the edges of the page.[46]

Despite the differences in style, the two became great friends because Raoul was the only colleague in Ghent Quetelet trusted to discuss Homer, Virgil and Horace. Raoul even did his best to dissuade Quetelet from the modern 'vaporous school which calls itself Romantic', suggesting instead that he would be much better served to write his reflections on Virgil's Georgics: 'These poems, the most perfect of their kind that remains from the ancients, are eminently proper to demonstrate the principles of sane literature and also the antidote to protect young people from the disease of this poor doctrine [of Romanticism]'.[47] After reading de Staël's *De l'Allemagne*, Quetelet offered a more balanced view of the relationship between the ancients and the moderns, but for the time being, he was content to learn as much about the ancients as possible from Raoul and continue his interest in the 'poor doctrine' of Romanticism.

Apart from their shared interest in classic literature, Quetelet and Raoul also became friends for more prosaic reasons: the hurried construction of the University of Ghent left little faculty housing available for the young faculty. In response, Raoul had purchased a large house that served as a commons for many of the younger (and poorer) professors. Quetelet rented out one of the rooms upon first arriving, and his description deserves an extended quotation for its portrait of a young artist and mathematics professor:

> This house, surrounded with gardens, later became the meeting-place for *un société choisie*. The most talented and distinguished young people were sure to be found in his home. There they received a welcome reception and an excellent consul for the direction of their studies: his library, his table, and even his pocketbook were placed at their disposal.[48]

Raoul served as more than a welcome host for Quetelet; he was the most ardent supporter and publisher of his poetry. To go only by Raoul's descriptions, Quetelet was the greatest poet since Horace. In an essay Raoul compared Quetelet's 'La Scalde et Lysis' favourably against all of the German-influenced 'Scandinavian literature', claiming that the 'melancholy and pain' found in it were exemplary of the superiority of *Midi* poetry. The 'felicity' of his pen was also to be found in 'La Châtellenies' and 'La Comtesse Ide', which appeared in 1819 and which Raoul described as 'fabulous'.[49] Quetelet appeared in the *Mercure belge* or *Annales Belgique* dozens of times, from his first effort, 'La Veillée des bardes', in 1817 to his long essay on the origins of Romanticism in 1823.

While Raoul's editorial positions certainly helped the young poet's artistic career, by 1819 the demands of Quetelet's dissertation and appointment in Brussels caused a severe decline in his writing. By then, even Raoul described Quetelet's poetry as 'only a diversion' from his scientific work, and for almost one calendar year between 1818 and 1819 Quetelet published nothing in order to focus on his qualifying exams and dissertation.[50] Always in good spirit Raoul never chastised his young friend for focusing too much on his career in the sciences, but in a later essay critical of Victor Hugo, Raoul worried about the loss of freedom for the artistic spirit at the hands of the sciences:

> The current thought which dominates is [Hugo's] theory. It is in a time of enlightenment and progress, like ours, where nothing remains stationary; but genius, for too long obstructed by the hindrances of precepts and rules, needs to be more independent. The arts of the imagination do not have less to harvest (*moissonner*) than the sciences in the fields of theories, inventions, and discoveries.[51]

Raoul's general concern reflected a very real problem for a generation of broad-minded thinkers in the first half of the nineteenth century who wished 'to harvest' both from the sciences and arts. It also reinforced the dichotomy mentioned by Bonald, d'Alembert, Pascal and Chateaubriand. As students like Quetelet and Dandelin began their careers, it became increasingly clear that 'the arts of the imagination' were not part of most job descriptions, especially government positions requiring technical training. Raoul placed the blame for this state of affairs on the Enlightenment, which 'denatured and corrupted literature'.[52] Raoul indeed seemed frustrated with both sides of the war of arts and sciences; he viewed *savants* as technically trained bureaucrats and the Romantics as groundless new writers influenced by the impoverished modern Germans. Quetelet, who was influenced by these 'corrupters' as much as he was by classical poetry, may have felt some unease at his mentor's strong language but never reflected on the contradictions of his friends and influences. In the last few years of their correspondence before Raoul's death in 1849, Quetelet and Raoul lost touch. While their exchanges reached a peak between 1820 and 1824, when Quetelet received an average of four (long) letters per year, the 1830s and 1840s saw only a handful in total.[53] By this point Quetelet was internationally known in the fields of statistics and astronomy, and the time for discussing Horace in Raoul's dosshouse in Ghent had long passed.

When in 1819 Quetelet left Ghent for a teaching position at the Athénée de Bruxelles it must have appeared a different world. Following the inclusion of the Belgian lowlands into the United Kingdom of the Netherlands, Brussels between 1815 and 1830 experienced the greatest cultural explosion since the reign of Charles V, attracting talent in the arts and sciences from across Europe. It was not yet the home for European refugees it would become during the revolutionary 1830s and '40s, when Marx and Engels among others made Brussels one of the most international cities in Europe, but it was far from the sleepy medieval city that had existed in the eighteenth century.[54] It also helped that Brussels was flush with new money, with the city set to 'eclipse its rival' Antwerp as the economic heart of the Low Countries by the 1830s.[55] Not only had the push to create new universities brought in foreign scholars to teach at the new universities, but Brussels also became a home to French exiles after the Bourbon Restoration, including the painter Jacques-Louis David, and the writers Arnault, Bury de Saint-Vincent and Berlier. Almost immediately upon arriving, Quetelet joined in the flourishing artistic scene, joining the Comité de lecture de théâtres royaux and the Société de littérature and sending a new round of poems back to Raoul for publication in the *Mercure belge*.[56] Though not on the cultural level of Paris, London or Weimar, all places where he would travel in the next decade, Brussels became, as Quetelet put it, 'a flourishing city [where] the letters, arts, and sciences were soaring'.[57]

If the above claim contained exaggeration, it was in describing the *sciences* in 1819 as 'soaring'. Quetelet complained later that Brussels and the Low Countries as a whole lacked the infrastructural and institutional organization found in France and England, and there was little in the way of scientific ideas or discussion in 1819. So impoverished was scientific thought at the time that Quetelet immediately gravitated to a friend of his mentor Raoul, the enigmatic philosopher and mathematician Charles Nieuport. The *Commandeur* de Nieuport (1747–1827), as he preferred to be called, was the political opposite of the young writer, but would become one of his greatest champions. Born of an aristocratic family in Paris in 1747, he had served a long and distinguished military career in France before the revolution forced him into exile. Upon arriving in Brussels, close to bankrupt, he found favour with William I and was given a sinecure as the court's liaison to the Institut de France and a pension that allowed him to concentrate on his research. Nieuport's treatment at the hands of the revolutionaries had left him 'in horror of liberal ideas', and his conservatism went so far that he joked with Quetelet in their first meeting that he had no problem with the freedom of the press. His only stipulation was that 'all newspapers were printed in Latin'.[58]

Despite these apparent conflicts, Nieuport was also the closest thing Brussels had to a famous mathematician, and his influence on Quetelet was immediate. Before the revolution, Nieuport had spent his considerable free time in mathematics and had worked with d'Alembert and Condorcet. Politics aside, he was Quetelet's only personal connection to the Enlightenment mathematicians. Though certainly outside of the *philosophe* circle, Nieuport's facility in maths earned him the respect of his liberal colleagues. In philosophy however, he sharply broke with Condillac's form of sensationalism, agreeing that there were no such thing as innate ideas, but believing Condillac's 'analysis' to be 'only composed of segregation'.[59] Instead, he proposed a holistic approach where mathematical investigation led to the discovery of Newtonian laws rather than Linnaean dissections. Seeing in the affable Quetelet a likely follower in the sciences, Nieuport encouraged his young disciple and introduced him to what existed of a scientific infrastructure in Brussels at the time.

Though the commander himself had no great love for literature, he indirectly helped Quetelet to enter the more active Brussels artistic community by a letter of introduction to the Baron de Reiffenberg. The eclectic philosopher and poet had been born in 1795 of a German family but educated in Mons, the capital of the industrial Belgian province of Hainaut. Like Dandelin and Raoul, Reiffenberg served in the military during the end of Napoleon's reign but temporarily retired from service in the relative peace that existed between 1815 and 1830. Though his appointment was as a philosophy professor at Louvain, he spent a good part of his time five miles west in Brussels with the refugee group headed by David and Arnault. Reiffenberg co-founded the *Mercure belge* with Raoul, and was something of a minor star himself, described by one commentator as speaking with a 'Voltairian sparkling language' and possessing a 'delicate, subtle and transcendent mind'.[60] Quetelet, possibly overstating the case, even claimed that his friend's 'arrival in Brussels was met with a genuine ovation'.[61]

Whatever the city's reaction, there was no question Quetelet was impressed by the baron, calling his first visit to his house 'unforgettable', similar to his appreciation of Raoul's stimulating household.

> What volubility of language, what grace and ease, as if words would not suffice for his energy! He was in continual movement, going, coming, and rapidly climbing up and down the shelves of his library. After about a half hour of this, I was stunned to the point that I could not think of a single thing to say. It was decided however that I would take an apartment neighbouring his, which would allow us to talk at any hour.[62]

The 'neighbouring apartment' was on the same floor, one level above a set of dungeons that had been used during the Brabant Revolution. In the spirit of Raoul's house, the interchangeable apartments of Quetelet and the baron became a gathering place for students and members of the French expatriate community including David, who lived there intermittently for two years. Describing the house almost thirty years later, Quetelet claimed there was never a bad moment between the two and that Reiffenberg's 'energy' and 'joy' kept his mind constantly at work. They would live in this strange amalgam of studiousness and boisterousness for three years, with only Quetelet's departure for Paris in 1823 breaking up the arrangement.

Unlike Dandelin and Raoul, Reiffenberg remained in constant contact with Quetelet during the 1830s, in large part because he was almost as well thought of in Belgian intellectual circles as Quetelet. Aside from his standing as a philosopher, Reiffenberg published dozens of poems and led the Société de littérature de Bruxelles 'at its most flourishing state', when it included Jouy, the novelist Lesbroussant and Quetelet's close colleague at the Académie royale the Baron de Stassart. But in 1837, in the wake of a scandal charging that Reiffenberg submitted articles he had appropriated from unpublished manuscripts to the Académie royale, his career collapsed. Quetelet, though critical of the severe condemnation Reiffenberg received at the time, made no apologies for his friend's dishonesty. Instead he viewed the incident as a tragedy that caused Reiffenberg's late works to be ignored.

In his last decade Reiffenberg did produce a staggering output of work, even by the standards of the time (Quetelet reports that in just his final fifteen days he composed eighty fables). Until the end he received constant support from both Quetelet and Raoul, but the unfulfilled promise of his early success weighed heavily on Reiffenberg. Blaming the baron for 'doing too many things at the same time',[63] Quetelet reflected at length on the many demands placed on his friend. He was a philosopher, an artist, and at times even a *savant*, but was never able to focus on one discipline for very long. Unlike Quetelet who used his university appointments to move up the ranks of the state bureaucracy, Reiffenberg only kept these positions to free up time for his diverse array of interests. Quetelet remarked on this by linking Reiffenberg's plagiarism scandal to the many demands he placed on himself. 'In letters like in arts', he wrote, 'nothing can be neglected or made without preparation'.[64] By contrast, in 1837, when Reiffenberg's many demands caused him to rely on cribbed notes for a presentation, Quetelet had already effectively winnowed his interests to creating Belgian scientific institutions. It was a path that would lead him to the fame for which his friend the eclectic baron seemed destined, and suggested that the eclectic life in practice was difficult to sustain.

All three of Quetelet's good friends from his early years were talented in both the arts and sciences, yet it was Quetelet who emerged as the brightest star from the small network of friends in Ghent and Brussels. Unlike his friends, however, who were either plagued by careerists they found in the new scientific organizations (Dandelin), outright distrustful of the natural sciences (Raoul) or burdened by too many competing interests (Reiffenberg), Quetelet seemed at ease taking up new responsibilities and transitioning from the life of a quasi-bohemian professor and artist to a well-respected man of science and institution builder. Though his artistic publication record exceeded that of Reiffenberg and Dandelin, he was able to leave poetry behind when it came to getting on with a career in the sciences. Part of the failure of his friends to attain similar status was certainly due to inherent talent, but much of it also resulted from the culture of a modernizing state which, by the 1820s, was looking for new kinds of men of science.

Conclusion

Despite the large number of poems, operas, comedies and essays Quetelet wrote, they take up but a small percentage of his writings collected at the Académie royale: an armful of folders among stacks of boxes. After leaving Ghent for a teaching position in Brussels in 1819, visiting Paris as a liaison for the construction of an observatory in 1823 and becoming editor of the continental journal *Correspondance mathématique et physique* in 1825, his literary production waned until there were only a few remaining works of poetry found in eulogies to his friends by the mid-1820s. In correspondence even, there were only hints that he retained an interest in the arts: the occasional mention to an old friend about the latest in the theatre or a brief quotation from a classic poet. The combined toll of state and practical demands, the loss of close friends and the divisive split between the sciences and the arts had all contributed to an erosion of the literary life he once pursued. The man described during the celebrations of his centennial in Brussels as a 'poète, littérateur, géomètre, physicien, astronome et statisticien' seems to have lost the first two designations before the age of thirty.

Yet while the demands of age and career certainly contributed to the decline in Quetelet's literary output, the amount and quality of the work refuse a simple dismissal. These were not mere diary notes by a self-indulgent adolescent, but rather fully formed poems in the style of Horace, translations of Ovid and Schiller, five-act operas with three-part songs, fifty-page epic poems and critical essays. Furthermore, the later themes Quetelet developed in his poems show that he did not simply enjoy writing as a pastime but that he was deeply immersed in questions of what constitutes art, what sacrifices are necessary to produce it and what the role of the artist can be in society. It does not take much of an interpretative leap to suggest that the struggles of Quetelet's many tragic protagonists, from the troubled author Dièse to the distressed Poet of 'La Poète et La Raison', were ones that were taking place inside his own mind as both his scientific research and his literary writing were coming to maturity. As a young man, examining the sacrifices necessary for art may have seemed an engaging thought experiment, but for a thirty-year-old with a wife and family, the questions became real and practical.

Whether Quetelet could have made a career in the arts rather than the sciences is an intriguing question yet obviously impossible to answer with certainty. He made a minor name for himself in Ghent and Brussels, but as he acknowledged in the 'Épître a Tollens', he was far from even the masters of Belgium. What is known, however, is that individuals like Dandelin and Reiffenberg who tried to pursue their interests in both science and art suffered professional and personal consequences. While some of the sacrifices necessary for art – poverty, devotion and suffering – had always existed, Quetelet also faced new challenges through the professionalization of the scientific disciplines, beginning with his dissertation and extending through to his apprenticeship in Paris. Coupled with his many obligations to bureaucratic reforms of scientific institutions, discussed in the next chapter, he no longer had the same network of friends who could help guide his artistic output. The friends, co-writers and editors who helped Quetelet compose and publish his verse and essays were replaced by advisers, mentors and administrators who helped him acquire the proper technical training in administering scientific organizations. Quetelet may not have been able to be a poet and librettist even had he given up a career in the sciences, but, as his own creative literature suggests, and the experiences of his friends confirm, he would have had great difficulty becoming one of the most successful nineteenth-century men of science without giving up the arts.

3 STOKING THE SACRED FIRE: THE ADMINISTRATION OF OBSERVATION IN THE UNITED KINGDOM OF THE NETHERLANDS, 1822–30

Building Belgian Science

If the war between the arts and the sciences diminished the opportunities for Quetelet's friends to continue in the spirit of the Enlightenment consensus, the same developments created the conditions under which Quetelet achieved his ultimate professional success. The modernizing European states of the nineteenth century in fact offered Quetelet an ideal context in which to promote his most enduring ideas, and it is no coincidence that a quantitative average man emerged at the same time these states were in the process of combining traditional state bureaucracy with new ideas on statistics to create a new fusion of empirical data and control.[1] Though Belgium in the early 1820s may have lacked the model integration of science and polity that had been built at the Institut de France by Turgot, Condorcet and Laplace, Quetelet was intent on modelling Belgian science on the institutions of his neighbours to the south. Based on his ability to create a new observatory, a government census, a thriving journal and a national commission on statistics, it was a success. The consequences of the intellectual war at the beginning of the century may have meant that Quetelet could no longer contemplate the heavens through both verse and empirical data, but the loss of a passion had enabled the beginning of a career.

His timing was fortuitous, as just about the time when his life as a Romantic bard seemed most in peril, the Catholic provinces of Belgium became part of the brief political entity known as the United Kingdom of the Netherlands. Though this almost forgotten kingdom lasted only 15 years after the Napoleonic Wars ended, the structure and ideals of the state had an important role in Quetelet's ideas after he stopped writing poetry. The polyglot territory included most of what today constitutes the modern states of the Netherlands and Belgium and was similar in size to the Spanish Netherlands that had existed prior to the Eighty

Years War and the 1648 Treaty of Westphalia. While there was some historical justification for the reunification, the new nation faced a number of divisions. Belgium had *not* gained its independence from Spain in 1648, was almost uniformly Catholic and remained under the control of the Hapsburgs until the French occupation in the early 1800s. The northern Netherlands had on the other hand won independence from Spain through a long and bitter struggle, based their independence on a rejection of Catholicism and had been an independent country and economic engine in Europe for centuries. Furthermore, Belgium itself was (and is) divided linguistically, with the northern region of Belgium (Flanders) speaking a minor variant of Dutch and the southern region of Belgium (Wallonie) speaking French. In the midst of all these divisions and instability, in a country still recoiling from the horrors of war and revolution, Quetelet began his plans to reform Belgian science.

Without question, the goals of the new kingdom were to bring stability and order to the nation. Ruled by William I of Holland, the explicit plan to bring this order was to 'Dutchify' Belgium in all matters economic, social and cultural.[2] The early move to reform the schools had led to the creation of the new universities in Ghent and Brussels. It also meant lower-level reform in the grammar schools of Belgium, schools described as 'influenced by a Voltairian attitude, anticlericalism, scepticism [and] religious indifference', and William made it central to his plan to remove the Catholic bishops from positions of prominence in the south.[3] Though concerned about French Catholicism and 'fanaticism', William also appreciated the technical focus of many of the schools that Napoleon had set up. Quetelet could therefore appeal both to the king's highest hopes and deepest fears in petitioning for reform in the institutions of Belgian science.

Though himself Catholic and not far removed from his reverence for Chateaubriand, Quetelet successfully navigated this new culture of state scientific practice in Belgium that had frustrated his friends, Dandelin in particular. By investigating his efforts in creating the Observatoire royal de Bruxelles (later, de Belgique), his teaching in the Athénée de Bruxelles and his reformation of the Académie royale, this chapter reveals Quetelet in fact to be a master of scientific administration who excelled in recruiting and directing large organizations of researchers. Though the result was not necessarily science as bureaucracy, in the sense of a Weberian hierarchy with delineated rules, formal procedures for advancement and 'channels of appeals',[4] or certainly anything like a Taylorist office, Quetelet modelled the structure of these institutions on the organization he saw in the bureaucrats in the government of the United Kingdom of the Netherlands. He explicitly cast administrators as models for his new science workers. In creating and reforming the observatory and the Académie royale in particular, Quetelet also cultivated a new kind of science worker to staff these institutions, a crucial development as he began to imagine a new science called *physique sociale*.

Model Men of State: Falck, Van Ewyck and the Creation of an Observatory

It is difficult to overstate the importance of the Observatoire royal de Bruxelles for the establishment of Belgium among the important scientific nations of Europe. As a later Belgian academy member put it 1996, Quetelet is still remembered in his home county for 'extracting us from mediocrity and elevating us to international recognition'.[5] As Quetelet would learn from visiting the Paris Observatory in 1823 (see Chapter Four), *national* observatories in particular were crucial for establishing nations as major scientific players in the eyes of both administrators and the public.[6] Observatories had been the most important physical manifestations of administrative science, not only conferring upon a nation a sense of status and legitimacy, but also serving as politically useful tools for rulers intent on modernization and progress. It was no different in Belgium, where Quetelet convinced William that an observatory would be central to the king's dream to remake the southern provinces of the United Kingdom of the Netherlands on the model of Flanders. The importance of the observatory was such that on the occasion of its one-hundredth anniversary, one commentator noted that the observatory was 'the premier scientific establishment' in the country and that, in large part due to Quetelet, 'after two centuries of intellectual ignorance, Belgium had regained a brilliant place among nations'.[7] Though the author of this somewhat optimistic reading of nineteenth-century Belgian science was the observatory librarian, the Observatoire royal was the primary institution for Quetelet's reform of Belgian science.

Figure 3.1: **Observatoire royal de Bruxelles.** Quetelet imagined the observatory in Brussels to be a symbol of order in the United Kingdom of the Netherlands. In making his case for the observatory to the king, he often stressed these political ideas over the scientific benefits. A. Quetelet, 'Plans et description des instruments de l'Observatoire Royal de Bruxelles', Annales de l'Observatoire Royal de Bruxelles, 11 (1857), pp. 3–18, plate 2. New York Public Library.

Quetelet made his first request in 1823, a year in which William would have welcomed the authority and stability of a national observatory. Almost immediately upon his acquisition of Belgium and the principality of Liège after Napoleon's defeat in 1815, the king planned a course of action that would in his mind elevate the southern provinces to the level of his native Holland. The first step was the formation of the three new universities in the south, including the University of Ghent, where Quetelet and his friends had been hired. As William made clear, one of the primary aims was to challenge the pedagogical authority of the Catholic Church. The restoration of the Bourbons in France had sparked fear in the north of the extension of a destabilizing and 'backward' regime into French-speaking Wallonie in particular. Indeed, it was feared that the new French minister Jules de Polignac had designs on re-establishing the 'traditional' borders of France to the Rhine, and the linguistic and religious overlap between France and the southern provinces made their political allegiance to William's fledgling state far from certain.[8] Further beneficial to Quetelet's case for a new observatory was the king's general temperament, with one historian describing the monarch as a famously sober thinker who 'grudged every hour spent on amusement'.

> He had no time for the study of any form of literature except works on trade and international law. It was said of him that he thought the most humble writer on political economy a greater man than Byron or Chateaubriand.[9]

Despite the fact that Quetelet had translated Byron just four years earlier, by 1823 William's administrative pragmatism meant he was the kind of man with whom Quetelet could work.

What is most notable about Quetelet's initial request to the king is that the traditional *scientific* benefits of an observatory – discovering comets, recording barometric pressure and providing accurate stellar maps – were not mentioned until the very end. Instead Quetelet began with an appeal to William's pride by lauding the history of science in Holland, home to the House of Orange and the model province to which the rest of the new Netherlands was to conform. It was Holland, Quetelet wrote, which 'gave the known world the first ideas of optical instruments which uncovered the mysteries of the skies' and was able 'to discover vast regions where the efforts of man had not yet penetrated'. They had also produced Christian Huygens ('the noble rival of Newton') who had 'extended the limits of astronomy by his ingenious work'.[10] Quetelet argued that by establishing an observatory in the southern provinces, Belgium could match the excellence of the astronomers from the king's birthplace.

More important than the legacy of great scientific discoveries – at least from the perspective of a ruler confronted with a possible Catholic enemy to the south – was the traditional strength of astronomy in confronting 'fanaticism'. This might explain why, in his request for money for a modern nineteenth-century observatory, Quetelet included praise for Philippe Laensberg (1561–1632), a Belgian astronomer and Calvinist minister who was removed from the latter position for maintaining the Copernican position that the sun was the fixed centre of the universe.[11] Quetelet portrayed Laensberg, who was born in Ghent but preached in Zeeland, as a martyr for knowledge and understanding in the face of oppression.[12] This astronomer, 'whose fate was that of Galileo', refused to abandon his position despite pressure from his church employers and fellow countrymen. According to Quetelet, Laensberg 'struggled courageously against the efforts of fanaticism, and when it became impossible to resist, he chose to leave rather than renounce his proper conviction'.[13]

What linked a seventeenth-century astronomer to the scientific merits of building an observatory in 1823? For Quetelet, the reference to Laensberg implied that an observatory would promote the 'sublime' field of astronomy which had historically represented a stable field of knowledge in the face of ignorance. Without explicitly mentioning the religious differences between the two halves of the United Kingdom of the Netherlands, Quetelet complained that while his compatriots in Belgium had been able to keep up with Holland in pure mathematics, progress in astronomy had slowed because of minimal government support. What went unstated, but which was surely appreciated by William, was that the previous Habsburg government of the south had been dominated by the local clergy of the Catholic Church. Just as in the case of Laensberg, astronomy in Belgium had been stagnated by religion. Not only had the ignorance of previous regimes hindered progress in astronomy, but 'above all' it was 'the rapid succession' and 'instability' of governments in the Low Countries that made progress so difficult.[14] William viewed instability as the major problem in his new lands, and Quetelet offered an observatory as a possible solution. The order and discipline required for observing the stars might be a model for the king's new territories.[15] After all, Quetelet wrote, William was a monarch who 'from the first years of his reign' had wanted 'the provinces of the south to enjoy the same advantages as the provinces of the north' and who had previously granted, with a 'truly royal munificence' three new universities.[16] While this was a good beginning, the south still lacked a scientific institution of the highest rank.

The observatory could then help William combat the two principle challenges to his regime: the potentially 'fanatical' religion to the south and the general disorder and instability which had plagued Belgium for centuries.[17] All that was needed was to promote astronomy, a discipline Quetelet described as 'a science which wishes to elevate itself and expand peacefully in silence; one which waits for generous and clear protection, and not for outrages and persecutions'.[18] The choice for William was clear: either he could help protect knowledge and agree to build the observatory, or he could abandon the southern provinces to 'outrages and persecutions'.

Quetelet's request could not have been put in better terms. He knew William's fear of the French and may have felt it himself, having experienced political chaos in his early life. The French rule of Belgium overlapped with Quetelet's first seventeen years of life and it has been claimed that during this period in Belgium 'there was only resentment, despair and apathy'.[19] Though William may have been interested in the ability to determine the position of celestial bodies to a greater exactitude, Quetelet wisely left the technical merits of an observatory to the end. He knew that the rationalism and order of the observatory was more palatable to the king than producing star catalogues. The order and stability of the stars, as viewed through the new telescope at the Brussels observatory, could portend the future possibility of order in human relations. It was a message that no self-consciously reforming leader could ignore. Though Quetelet was not officially named director until five years later, he was charged in the summer of 1823 with putting together the equipment and planning for the new Observatoire royal de Bruxelles.

Though William had approved the plan to build the observatory, Quetelet was supported in his efforts by a career bureaucrat who may have had the most influence on Quetelet's vision of scientific research: the minister of education, Antoine Reinhard Falck (1776–1843). Quetelet had first come to Falck's attention in 1820 at a meeting of the Académie royale, when Falck had been impressed by the young poet's 'Éloge de Gréty', published that year in Raoul's *Mercure belge*.[20] Years later, when Quetelet first proposed the idea of an observatory to Falck, he stressed the importance for the future of the country. Claiming that he 'dared to speak' to a figure as important to Falck, Quetelet reported that Falck had 'the kindness to listen to the advantages [of an observatory] ... both in the sciences in general and the country in particular'. Quetelet, admitting the 'temerity of his request' – having never set foot in an observatory and now proposing to build one in the centre of Brussels – found that Falck was disposed to listen to the plans for 'the country in particular'. Falck agreed to meet with the young Belgian again, beginning a formative relationship that aided Quetelet's plan to remake Belgian science.[21]

A lifelong administrator, Falck had served in numerous positions in the Dutch government, from secretary of the ministry of foreign affairs to minister of commerce, colonies, public instruction and beaux-arts for the United Kingdom of the Netherlands, eventually becoming ambassador to England. At each level, Falck was a dedicated and able servant to William's plan to reform the southern provinces, beginning with the creation of the universities, which enabled Quetelet's first professional job, and the reorganization of the Académie royale in Brussels. Though the *ministre* had scientific ambitions in his youth, Quetelet described Falck as the ideal bureaucrat, one who 'above all, was devoted to his country and ready to sacrifice himself for these interests'.[22] Falck also shared with William a deep distrust of the French, and Quetelet similarly appealed to the political utility of the observatory in maintaining the loyalty of the southern provinces. As a young man, Falck had fought against Napoleon's invading army and felt that the temporary annexation of Holland to the French Empire had been the 'total ruin' of his country. In short, he was the perfect man to oversee the observatory project. As Quetelet assured William, 'M. Falck has shown himself to be the most ardent partisan of the formation of an observatory'.[23]

Though Quetelet's social physics in the 1820s and '30s owed a great debt to quantitative thinkers like Laplace, Gauss and Condorcet, he was never far from having an equal estimation of the importance of state administrators. On Falck, for example, he made the surprising claim to William that this important, but hardly famous, advisor was 'in some sense the personification of a great era in the history of Holland'. Perhaps Quetelet was just offering the standard breathless praise of an *éloge*, but he went further to praise all bureaucrats, claiming that 'Men of state are a little like metals: they are not estimated and appreciated until after having passed several centuries under the ground'.[24] The historical profession has not yet justified this axiom in Falck's case, but the claim itself provides evidence of Quetelet's true appreciation of the work of government officials. Rather than viewing the bureaucratic aspects of science as a simple necessity for advancing his scientific career, it seems Quetelet may have appreciated them more than the theoretical and empirical activities of scientific research.

Not everyone agreed. Quetelet had composed his letter to the king in 1823, but delays ensued. He had been sent to Paris that summer to investigate astronomical instruments, but the observatory was not formally announced until 1826 and Quetelet was not named director until 1828. During this time he encountered obstacles to his plan, as Falck was named ambassador to England in 1824 and his replacement Daniel-Jacob Van Ewyck (1786–1858) seemed less receptive to Quetelet's appeals. Though Quetelet had expressed hope for the new *ministre de l'instruction* in his appeal to William, calling him a 'man very interested in the secret of science', the new official proved problematic.[25] Trained in mathematics and physics, Van Ewyck deflected many of the political arguments and challenged Quetelet on the *scientific* uses of the observatory in a way that neither William nor Falck could. Quetelet's response – frustrated, defensive and elusive – showed that by the middle of the 1820s Quetelet still had more success with political arguments than scientific ones, and that the observatory had a stronger claim to advancing the interests of a stable Belgium than to contributing to international science.

In his first letter to Quetelet, Van Ewyck began by agreeing that the observatory would be a 'great glory to the kingdom' and praising Quetelet for his dedication to the state. However, he followed with that most ominous of lines:

> But things can be seen from another point of view ... What great discoveries can one reasonably hope to make in astronomy? Can one claim to make a better catalogue of the fixed stars than we currently have? The parallax remains [a problem], but could this be resolved with new instruments? For these reasons, I do not know if science would profit much from the erection of a great observatory.[26]

Quetelet actually did hope to make a better 'catalogue of fixed stars', but the criticism that 'science would [not] profit much the erection of a great observatory' was backed by the common assumption of the day that all of the great discoveries had been made in astronomy.[27] As J. A. Bennett has written, even for noted astronomers like John Herschel, 'utility alone could scarcely have excused the duplication of effort' brought about by more and more observatories.[28] Quetelet had offered a few scientific benefits to William, but he refused to respond directly to Van Ewyck's question of new discoveries. Instead he fired off an angry letter to Falck complaining about the unhelpful administration of his replacement. Quetelet's response to the question of discoveries was elusive: 'to indicate them in advance is to have already made them'. Quetelet repeated this defence several times ('I can never leave this subject'), but while a clever response, it was not the kind of answer designed to win over a sceptic such as Van Ewyck.[29]

The problems with securing financial support persisted even as Quetelet travelled to Paris and Germany to learn the details of administering an observatory. New observatory equipment was particularly expensive in the 1820s, as the nineteenth century had seen a great enthusiasm for larger telescopes equipped with achromatic lenses after William Herschel's successes.[30] At the Paris Observatory in the late eighteenth century, Jean Dominique Cassini (Cassini IV) had even tried to replace the large observing instruments – 'sectors, quadrants and circles' – with expensive single-cast models that did not need to be assembled in pieces.[31] The 'celebrated parallactic telescopes' in particular were the latest sensation in observatories, with one excited report describing how the lens had increased in size from a seven-inch diameter to a possible '*eighteen inches* in diameter', costing an estimated £9,200. In 1827 Otto Struve's Dorpat Observatory acquired a much more reasonably sized telescope for the discounted price of £950 (full price – £1,300).[32] Quetelet reported two years before that the Paris Observatory had recently received a similarly sized *lunette parallactique*, a clear sign that other observatories were pulling ahead before the Brussels Observatory could even start.[33] As will be seen in Chapter Four, Quetelet spent much of the 'Astronomy' section of his continental journal, *Correspondance mathématique physique*, chronicling the great acquisitions of other European observatories rather than reporting actual empirical data from these observatories, a decision that mirrored his interactions with administrators in the United Kingdom of the Netherlands.

Rather than explain the potential benefits of such expensive equipment, Quetelet instead chose to share his doubts about Van Ewyck with Falck. He claimed the current minister's arguments were 'weak', but still failed to provide Van Ewyck with examples of the specific scientific discoveries that could be made with a new observatory.[34] Instead, he relied on testimonials from the important contacts he had made during his travels and correspondence, writing that 'MM. Poisson and Laplace have said to me that they take a genuine interest in seeing the creation of an observatory in Brussels', and that Quetelet's 'friend' John Herschel had expressed support in a recent meeting.[35] By 1826, he was complaining that he had even declined a position at a university while he waited: 'I have burned all my ships; my fate is to await an observatory in Brussels'.[36] This last letter, sent 17 June 1826, was unnecessary however: cross-posted just a few days earlier from Amsterdam was a copy of the Royal Order, signed by William on 6 June, officially declaring support for Quetelet's project. The king and Falck had apparently won out over the sceptical Van Ewyck. Even without a sufficient scientific rationale to convince Van Ewyck, the observatory was to begin in two years.

It would have appeared a good time to be confident in the future of the observatory in a reformed Low Countries, and there were few signs of the short future for William's unified Netherlands.[37] As in so many other plans for the development of the sciences in Belgium, the Revolution of 1830 caused a delay in the much-anticipated progress. Indeed, it was not until 1835 – a full twelve years after the initial proposal – that the large telescopes and circles finally arrived from Paris and London.[38] Not until then could Quetelet collect independent data on stellar movement and Belgium take its place among the great scientific nations. It would be too late, however, to save William's kingdom.

During this tumultuous period Quetelet was forced to find other ways to occupy his time and, given the immense amount of work he completed between 1823 and 1835, it might be argued that Van Ewyck's objections, construction delays, and even the Belgian Revolution itself, prodded his entry into statistics. Certainly the construction of the observatory necessitated his trip to Paris, where he first encountered probability theory, but the delays also allowed him to teach, travel, edit the *Correspondance mathématique et physique* and present his findings at the Académie royale. All of these activities would contribute to Quetelet's *physique sociale* from 1823 to 1835 and it was during this period when he made his most lasting contributions in statistics and social physics. In fact, once Quetelet was able to take up full control of the new operational observatory in 1835, he was content to exchange his new data with an international network of astronomers and to establish more lasting government institutions, leaving the tasks of interpretation, experiment and analysis to others. In 1826, Alexis Bouvard, the director of the Observatoire de Paris, anticipated the course of Quetelet's career once he became head of such a major institution, repeating in a letter to Quetelet what he had told his friends in Paris: 'When the observatory is perfectly organized, M. Quetelet must not occupy himself with other things foreign to astronomy'. Bouvard warned that 'One must be always occupied in this science', and that attention to other disciplines led to 'neglecting the principle field' of astronomy.[39] Once again the possibilities for intellectual life were bounded by institutional concerns.

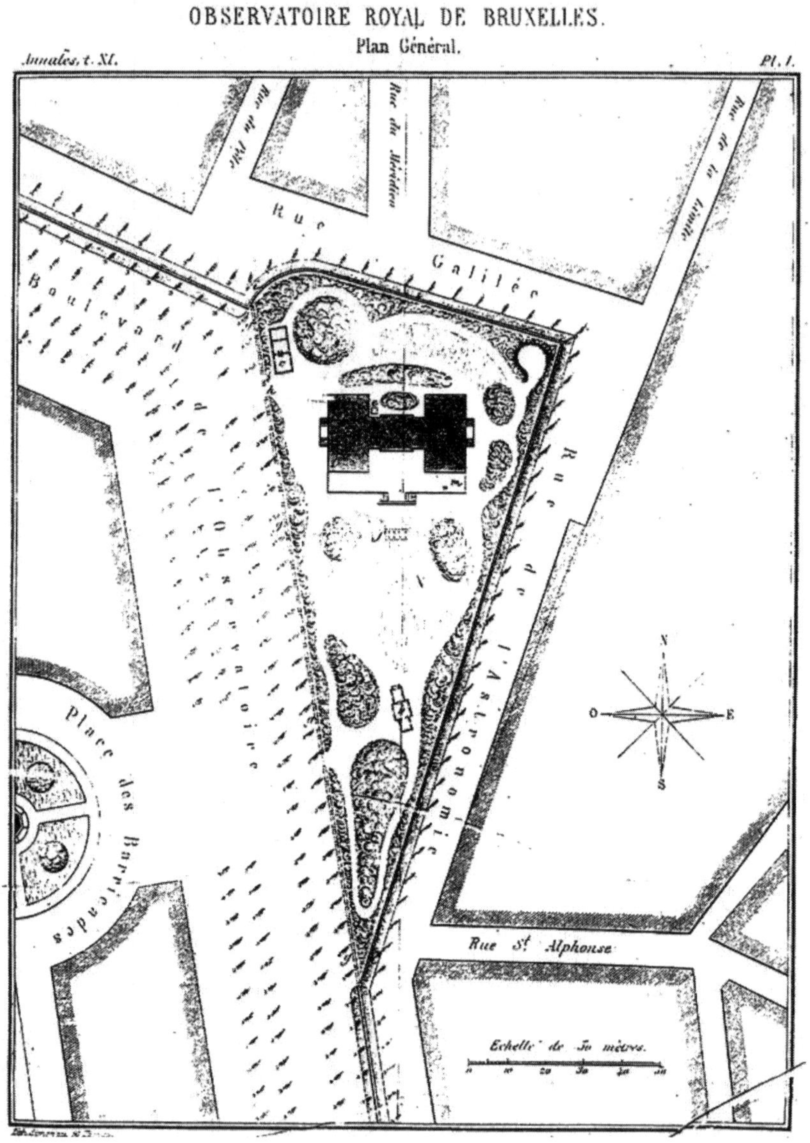

Figure 3.2: Observatoire royal de Bruxelles, plan général. Though Quetelet had envisioned an observatory as early as 1823, it was not fully completed until 1835. In the interim, he collected much of the data that would form the basis of physique sociale. A. Quetelet, 'Plans et description des instruments de l'Observatoire Royal de Bruxelles', Annales de l'Observatoire Royal de Bruxelles, 11 (1857), pp. 3–18, plate 1. New York Public Library.

Fortunately for later practitioners of statistics and the social sciences, the delays of Van Ewyck and the Belgian Revolution allowed Quetelet plenty of time to explore fields 'foreign to astronomy'. It is possible that had the observatory been finished earlier, social physics would never have been developed. Indeed, writing near the end of his life, Quetelet made the astonishing claim that the research that formed the basis of *Sur l'homme* 'had allowed [him] some distraction in the middle of my work in astronomy'.[40] Though a remark possibly made in an effort at humility, it had elements of truth. In a later letter to Bouvard, he complained that 'in the universities one can only practice the principles of astronomy and cannot obtain a rigorous precision in observations'.[41] Burdened with only the principles of astronomy, without an observatory, precision in numbers had to be found though examination of population and criminal records – the largest sources of easily obtainable information. However, once he was secured atop the highest elevation in Brussels, Quetelet was free to pursue his two favourite activities – large data collection and the construction of scientific institutions – and leave the 'practice of principles' to others.[42]

Popular Astronomy?

While the observatory was a symbol of order imposed from on high, Quetelet also had a plan to inspire the Belgian population from below: a series of public lectures on astronomy for the Athénée de Bruxelles. While Quetelet had taught physics, mathematics and astronomy at Ghent, those classes were intended as technical training for a new class of investigators and were therefore unintelligible to the general population. The *athénée* would be different. In order to begin William's plan to 'Dutchify' Belgium, the most 'obvious' approach was to remedy education, and in 1825, all religious boarding schools without a state license were forced to close. To replace these *petits séminaires*, a series of new schools opened throughout the southern provinces. The new *athénées*, the ones inspired by 'Voltairian attitudes' and 'anticlerical scepticism', were designed to transform the next Belgian generation into a practical body, one that could compete with the other great nations of Europe. Quetelet was more than willing to give his assistance.[43]

Quetelet's two courses at the *athénée* were well received and the published versions of these lectures – *Astronomie élémentaire* (1826) and *Astronomie populaire* (1827) – were later used for instruction in the Royal Colleges of Paris and Versailles, among others.[44] For those already too old to be brought into the new schools and who had attended the religious *petits séminaires*, Quetelet offered a basic introduction into how astronomy worked, providing summaries of what was known about the movement and positioning of stars, planets, the sun and moon. Quetelet wisely avoided much of the technical language and equations found in so many of his other works at this time. As he realized, astronomy was a 'science essentially founded on calculus', and he was working with adults who lacked quantitative aptitude.

I have tried in exposing the principle results of the work of geometers, to make understood a way for others to elevate themselves to the genius of man, by preaching the laws [of astronomy] founded through an assiduous observation and by a thorough knowledge of mathematics. Scholars cannot ignore how difficult it is to place at the door of people the results of a science essentially founded on calculus.[45]

How then to teach astronomy without mathematics? For Quetelet the answer was to rely on the most important interests of his youth: Romantic and classical literature, including poetical and mythical references. Though Quetelet believed that the well-ordered and rational sky inspired stability in a nation, and chose this message in his appeal to the king, he opted not to emphasize order in his classes for the *athénée*. Instead, Quetelet's lectures appealed to the beauty of the heavens rather than their importance in the hierarchy of the sciences.

Quetelet's longer work, *Astronomie élémentaire*, was filled with classical and literary allusions. After a brief survey on how astronomical data was accumulated – including another plea for a Belgian observatory – Quetelet told the mythical stories of the most famous constellations.[46] In describing the positions of *Ursa Minor* and *Ursa Major*, for example, Quetelet chose not to begin with their importance in navigation, or with their discovery by astronomers, but with the myth of Callisto and Arcus from Ovid's *Metamorphoses*. After Juno, 'irritated by the lies of her husband', had turned Callisto into a bear, Quetelet explained to his students how Jupiter saved the nymph and her son by 'placing them in the sky'. In a final fit of revenge, however, Juno sent the great dragon (the constellation *Draco*) to guard the mother and son. Quetelet of course was not offering an actual account of the creation of the stars – lest there be too much confusion, he made sure to be explicit that he was referring to 'ancient beliefs' – but the placement and prominence of the description went further than providing colourful background.[47] The pattern of ancient references continued with an epigraph from Virgil's *Georgics* prior to the discussion of Orion, more myths from Ovid and a chivalric poem by the eighteenth-century Gantois poet Charles Nieulant, all meant to provide a base for the study of astronomy beyond its practical applications.[48] Quetelet knew the technical details of astronomy quite well, but he offered his students myth and poetry in his initial appeal.

Yet in combining myth and science Quetelet was careful not to wander into astrology, an always dangerous detour when dealing with popular astronomy in the nineteenth century. After the long section on the mythical descriptions of the constellations, Quetelet scuttled any speculation in the direction of that 'chimerical science'. The confrontation of astronomy and astrology was cast explicitly in Enlightenment language with the former, 'like all other *connaissances humaines*', having to 'elevate itself above the imposters that had abused them to take part in credulity and ignorance'.[49] It was the 'spreading of enlightenment', Quetelet believed, that had 'tended little by little to shed light on men and deliver them from their errors and their prejudices'.[50] Though Quetelet admitted the importance of astrology in inspiring the discoveries of the past – specifically the advances of the Italian mathematician Gerolamo Cardano who studied the stars for signs on the life of Jesus – he refused to accord these speculations the same status as classical myths. While the myths initially inspired speculation of the heavens, the focus on astrology had amounted to little more than a waste of talent and resources, and it was a great 'regret' that so many learned men had sacrificed their efforts to the 'received ideas' of their time. Cardano, for example, had been so interested in gambling and astrology that he had almost been forgotten by Quetelet's time.[51] Only through the heroic efforts of Copernicus and Galileo, Quetelet claimed, could this 'debasement' of astronomy be overcome.

In his other work, *Astronomie populaire*, Quetelet dispensed with an explanation of the practice of astronomy – in fact almost all technical information – and focused instead on a straightforward listing of the principle stars, planets and constellations. Each section also included a set of questions at the end – 'What is the number of stars in the universe?', 'What is Astronomy?' and 'What Makes a Star a Star?' – that were similar to a standard elementary school textbook. The descriptions of constellations and stars themselves included no classical myths or poetry, and constellations received their proper names. Far from the mix of literary and scientific explanations from the *Astronomie élémentaire*, this volume featured a scaled-down presentation of facts alone.

As a dry chronicle of what astronomers knew about the heavens, the presentation was in direct contrast with the intentions Quetelet expressed in the introductions to both works. The goal in each case was to explain to the public the highlights of a field that, while capable of fascinating discoveries, was limited in practice to all but the most fortunate observers. As Quetelet learned in Paris and Germany, astronomy practised in an observatory was significantly different from what one learned as a student. The technology included in observatories had advanced to the point that the small telescopes that could be found at most universities were incapable of keeping up with the tasks set out by the international network of astronomers. The lengths to which Quetelet went to make astronomy meaningful indicated just how far the public was from an understanding of the discipline's technical means.[52]

Quetelet believed astronomy had become one of the first fields to require not only time and money but government support and international connections. A number of independent amateur astronomers remained, but Quetelet envisioned institutional affiliation as the most likely path to success in the discipline. In the introduction to *Astronomie populaire*, there is even an admission that the astronomy of his youth – independent, abstract, gentlemanly and literary – was gone, not soon to return. In a paragraph meant to explain why his field had such a difficult time attracting public interest, Quetelet explained that 'today [astronomy] seems to have become the exclusive domain of a few *savants* who, by their fortune or by the support of their government protector, can procure the necessary instruments to work with great success'.[53] Quetelet of course knew all too well the importance of a 'government protector', as not only was he awaiting the William's assistance, but many of the most notable observatories he visited on the continent – Paris, Göttingen and Altona – were constructed with government funds.

Ironically given the lecture title – 'Popular Astronomy' – Quetelet also believed that the construction of larger observatories was creating apathy for the science among the general public. Later marvels of public spectacle like the Berlin Urania (named after the Greek muse for astronomy) had sought to have a separate 'public telescope' and to combine 'scientific showmanship' and 'the workings of a professionally equipped observatory'.[54] But Quetelet ironically feared that the state had appropriated the everyday uses of astronomy from the people. For example, he claimed that before uniform star catalogues, people had kept their own maps for travel as well as calendars. Because people needed to know the basic positions of most stars, they had a practical interest in learning at least the outline of astronomy. But in the eighteenth century standardized maps and calendars had replaced private ones, lessening the general appeal of amateur astronomy: 'Since people have ceased to make calendars for themselves, and it was now done for them ... they have ceased to be able to even appreciate the service that is given them'.[55]

In light of this comment, the paradox of Quetelet's mission to popularize astronomy might become apparent. On the one hand he saw astronomy as a metaphor for order and stability, one that would contribute to a more stable political order were it embraced by the people. Yet on the other hand, it was the professional science of observatory astronomy that was causing the general interest in astronomy to decline. Quetelet certainly knew enough mythology and literature to keep things interesting in his classes, but he admitted that the great stories of the constellations had no real role to play in nineteenth-century astronomy. He also believed that the more governments succeeded in controlling the production of the discipline, the more they undercut its appeal to individuals. Both of Quetelet's popular books on astronomy tried to correct the decline in the public interest, but they were bound to be counterproductive, as he continued to argue for the creation of the same expensive and exclusive institutions that had created the problem in the first place.

The 'Rare Men' of the Académie Royale

With the observatory and his courses at the Athénée de Bruxelles, Quetelet did much to build a national scientific reputation, yet he was not responsible for *every* Belgian institution of science.[56] There was at least one major institution in Belgium that predated his arrival: the Académie royale des sciences, des lettres et des beaux-arts de Belgique. This organization, originally known as the Académie impériale et royale des sciences et des belles-lettres de Bruxelles, had been at the heart of the fledgling intellectual life in Belgium in the late eighteenth century but was close to moribund when Quetelet first petitioned for membership in 1820. The official historian of the institution described the primary activities of the *académie* during the years following French occupation as 'slow and routine ... leisurely meetings'.[57] Yet the speed with which Quetelet redirected the academy's purpose was remarkable. By the time he stepped down from his position as secrétaire perpétuel (more than fifty years after first being named a member), the *académie* had been transformed from a relaxed institution where old men discussed philosophy, literature and science into a youthful and organized body of professional researchers.

Quetelet reformed the institution in three significant ways. First, he sought out younger men of science and writers to supplant the informal atmosphere of distinguished gentleman and amateur researchers. There were a few good scientific minds remaining from before the French occupation, but sixteen years of educational neglect had resulted in almost an entire generation passing through Belgian schools with little training in the sciences. The period from 1815 to 1820 however saw a profusion of university-educated talent, and Quetelet viewed the *académie* as the natural next step in the process of creating professional researchers. Secondly, Quetelet worked to expand the number of foreign members, soliciting papers and research from France, England and Germany to bolster the meagre output in the *académie*'s *Annuaire*. Fearful of foreign influence, the *académie* had previously prohibited contact with foreign scholars.[58] Obviously Quetelet's plan to coordinate astronomical observations would require correspondents from further away than Liège and Maastricht, and under his direction, the class of sciences came to include dozens of the most famous scientists in Europe, mostly individuals that Quetelet had met with personally while travelling. Lastly Quetelet sought to promote members of the academy into the ranks of government administration. The *académie* never became a copy of the École polytechnique but it provided the government with a number of scientists and engineers. William had strongly supported a reinvigorated *académie* for just this reason, and in 1818 that great champion of the observatory and Quetelet's model 'man of state' Falck was named the first honorary member.[59] Through these three sets of reforms, Quetelet succeeded in creating a revolution in Belgian science. The revolution was not in the discoveries, however, but in the practice of science, as the old gentleman researchers gave way to dedicated workers in the mould of state administrators like Falck.

Upon arrival in 1820 Quetelet faced formidable obstacles in overcoming the malaise that had overtaken the *académie*. As he recalled dryly from this period: 'The first meetings of our academy often left much to be desired'.[60] Transportation had made it difficult for scholars outside of Brussels to make regular meetings, and the publication records of the *Annuaire* had been spotty. Indeed, the secrétaire perpétuel Charles van Hulthem lived in Ghent and spent the majority of his time alone in his study working on restoring precious manuscripts.[61] Both his interests and personality seemed more suited to the eighteenth-century clubs or salon life than directing an organization that William believed would 'reinvigorate the intellectual life' of the country.[62] As Quetelet described the meetings led by van Hulthem, they lacked order, though 'they were not without their charms, full of pleasant chats and scholarly discussion'.[63]

Fortunately, Quetelet was not alone in wanting to move the institution beyond 'pleasant chats'. His friend the Commandeur de Nieuport had been a member of the old Académie impériale and was frustrated at the direction of the *académie* under van Hulthem. In 1820, Nieuport was the most eminent mind in Belgium, and under his pressure, van Hulthem agreed to return to his study and leave the *académie* in better hands. Quetelet could, of course, also count on the support of Falck, a tireless supporter in almost all his projects, as well as King William. More surprising perhaps was the support of Van Ewyck, Quetelet's nemesis in the observatory project. But the *académie*, as an institution directed towards the actual production of scientific knowledge rather than a symbol of political order, was a more natural fit for Van Ewyck's vision of state-supported science than an observatory that could not promise any new discoveries. In describing Van Ewyck years later, Quetelet expressed his gratitude in language far different from that which he had used in his angry letters to Falck. 'One appreciates the care of ... Van Ewyck, a man well-versed in knowledge of interest to the Académie, and one on whom we could always count on for his intelligent support'.[64] With the help of a modernizing king, the leading scientific mind in the country in Nieuport and two of the most powerful men in government in Falck and Van Ewyck, Quetelet began his reforms in earnest.

On 1 February 1820 he was elected to the class of sciences in a unanimous vote. Just a few months later, Louis Dewez was named the new secrétaire perpétuel, pacifying Nieuport and marking the beginning of a new era for the *académie*. As Quetelet described him, Dewez was the perfect man for the job: 'there was scarcely a more exact man or more religious observer of his duties'. A historian of ancient literature in his youth, Dewez had 'turned toward the administrative functions' late in life and shared with both Quetelet and Nieuport a desire to reform the institution.[65] Though Quetelet did not take Dewez's place as the head of the academy until 1834, the old historian proved a useful proxy for his ideas.

To say that the name 'Quetelet' appeared often in the transactions of the Académie royale during this time would be a significant understatement. Rather, the energetic and affable reformer dominated the proceedings, speaking far more frequently than any other member during the 1820s and showing up nearly as often as the presumed leader Dewez. In just the period from January of 1822 until June of 1825, a time that overlapped with his travels to Paris and his classes at the *athénée*, Quetelet presented a memoir by Dandelin, proposed his friend for membership, reported on an survey expedition to Han, presented a paper on *conchoids circulaire*, proposed his friend Reiffenberg for membership, lectured on comets, published a note on caustics, suggested the membership of Gaspard Pagani, published a paper graphically describing the orbit of planets and, finally, presented to the academy his first paper on population data.[66] No other member from this time comes close to making this many appearances, despite Quetelet's relatively recent membership and his teaching obligations. Serving as a guide and example to his friends and colleagues, Quetelet practised the same self-devotion towards organization and scientific research that he preached to his young colleagues.

As part of the movement to modernize, Quetelet recruited a number of younger researchers, including many of his fellow faculty members from Ghent, sponsoring the membership of Reiffenberg and Dandelin in 1822 and 1823. But while his old friends had difficulty in adapting to many of the new requirements for science in Belgium – large group projects, centralized observation and impersonal assessment – Quetelet also found many younger members who shared his interests: François Cauchy, an engineer from Namur; Jacques Crahay, a meteorologist who would later conduct research at the observatory; Antoine Belpaire, a participant in the founding of the Commission centrale de statistique; and Pierre Simons, a young engineer who was slated to direct the Belgian delegation to the construction of the Panama Canal.[67] Each of these men, all under thirty-five, would go on to take important positions in Belgian science and represented the first generation of the 'new man' that Quetelet saw as the future of the country's development. By the 1830s, the old guard, 'too aged to understand the transformation in spirit' as one writer put it, had mostly followed van Hulthem to their respective private studies, leaving the *académie* to its most active participant.[68]

Who were these men *not* 'too aged to understand the transformation in spirit'? Quetelet's long life afforded him the opportunity to reflect on the lives of many of his most dedicated servants in the form of *éloges* and his mammoth *Sciences mathématique et physique chez les belges au XIXe siècle*, and through these reflections it is possible to get a sense of the talents Quetelet found useful in his new recruits. The most common word Quetelet used was *consacré*, meaning most likely a dedication or devotion to the task at hand. Academy members like Pagani for example 'dedicated his fortune and existence' to the sciences, while Antoine Belpaire 'sacrificed his pride' in order to make himself 'useful'.[69] Such words echoed the praise Quetelet had for his mentors in administrative positions, from the minister Falck to the director of the Paris Observatory Alexis Bouvard (see Chapter 4). They fit well with his vision of how the sciences needed to operate in the future and provided the template for all the men of Belgian science.

Though Quetelet trained dozens of the people he called 'rare men'[70] to take up positions in the sciences of the United Kingdom of the Netherlands, a few examples may suffice. François Cauchy (1795–1842), the engineer from Namur, was another *académie* member who 'dedicated his entire life' to larger projects, including directing a mining survey in Luxembourg for William's industrial expansion.[71]

Cauchy seemed the perfect fit for administrative science, moving from director positions to teaching mineralogy and 'accept[ing] each [position] as one accepts a new duty to fill'.[72] Cauchy rarely if ever published any new research and seemed content to advance to higher and higher positions in the state bureaucracy, eventually becoming director of the Carte minière de Belgique, an enormous project that sought to survey for potential exploitation the entirety of the king's land in the Catholic Low Countries. His official notice even claimed that Cauchy's 'administrative career' should be considered 'a model' for Belgian science.[73] Jacques Crahay too was an assistant at Quetelet's observatory and used his 'patient and organizing mind' to record atmospheric and meteorological data day after day.[74] Other *académie* members had an 'austere character', 'perfect integrity' or had a love for science 'that absorbed all of their time'.[75] It is not hard to imagine that these new 'rare' men of science were the same figures that Quetelet's friend Dandelin had complained about during his dreary trips to make yet another mineral survey – men interested in career advancement and the technical responsibilities of their work but little else. So too were the many mining missions the kinds of activity that Quetelet's poetic friends Raoul and Reiffenberg had railed at in Ghent and Brussels.

Though Quetelet's use of dedication and sacrifice was largely meant as metaphorical, many of the academy's rare men can be said to have literally suffered for their work. Soon after starting work for the Carte minière de Belgique, Cauchy became ill from a 'malady of the chest'.[76] A career dedicated to exploring mines had left him weak and incapacitated by his early thirties. Quetelet even speculated that the work on mining surveys had 'perhaps accelerated his premature death' at the age of forty-seven. What was called 'the summation' of the young engineer's 'life's work' had, Quetelet suggested, actually killed him.[77] Another dedicated young *académie* member who suffered physical harm in pursuit of science was Pierre Simons (1797–1843), a brilliant surveyor who would have been the Belgian delegate to the Panama Canal at age thirty-three if not for the Belgian Revolution. Simons had 'dedicated himself to a career in public works' since the age of eighteen, a dedication that took its toll on the young surveyor. Quetelet noted that Simons was so attached to his work that he was 'trapped in the middle of his papers' and 'only two or three of his friends would have known of his existence had he not had to leave his office'. This devotion had caused a number of health problems for Simons, who before even turning thirty was suffering from 'an excess of fatigue' due to his 'painful duties'. His health had been 'visibly affected' by research, and by his forties Quetelet described him as in a 'sickly state'. If the possibility of martyrdom for the canals and mines of Belgium were not yet clear to his readers and listeners, Quetelet noted that even in his last days 'the *feu sacré* had not yet gone out' of Simons.[78]

During this same time Quetelet was tending to the 'sacred fire' of domestic science workers, he also took advantage of the loosening of organizational policy on foreign members to recruit able *savants* from France, Germany and England. In all, thirty-six foreign 'correspondents' were elected during the period between

1816 and 1830, many in direct response to Quetelet's personal requests.[79] Though many would have joined anyway, it is notable that almost all of the most famous foreign members – André-Marie Ampère, Alexis Bouvard, John Herschel, Samuel Brown, Alexander Humboldt and Edward Sabine – were men that Quetelet had either personally met or who had contributed to his *Correspondance mathématique et physique*. Not only did foreign members add prestige to the organization, they allowed for international contacts on scientific research which required observations over a large geographic area. Though at this time it was far from becoming the home of international science that Quetelet had envisioned, the *académie* had at least succeeded in undoing the long intellectual isolation that had persisted under Hapsburg and French domination.

Once the *académie* rebuilt its domestic membership and attracted the notice and research of foreign members, it remained to integrate the membership into the growing state-sponsored research projects. From its reorganization in 1815, the *académie* had been dedicated in part to the production of useful knowledge and practical technologies. Article Twenty of its charter made this suggestion explicit:

> The Academy will examine, when the Government ordains it, projects concerning new factories, manufactured products, machines or the perfection of some useful arts. It will at the same time explain the types of advantages which can be found in the executions of these projects.[80]

The commitment to practical technologies and discoveries was not surprising. If William was going to succeed in integrating the southern provinces into his kingdom, he needed to industrialize the new provinces through projects like mines, railroads and canals. Not only were the educational reforms of new universities needed to break the hold of the Catholic Church, but new industries and technologies were required to undo the traditional worker cooperatives and guilds. In incorporating technological and practical advances from the sciences when 'the Government ordains it', the *académie* became a crucial part of reforming the Belgian economy and countryside.

A large professional and internationally linked institution of science, composed of selfless and dedicated researchers, would remain an *idée fixe*. In a later speech given to the assembled classes from the *académie* in 1852, Quetelet outlined the historical development of European academies from their earliest days to the present, leaving no doubt that he expected the growth of scientific institutions to continue in the form of research he had begun at the *académie* in 1820. As Quetelet explained, while the first academies of the seventeenth century had been founded to help individuals 'in their works and to second their efforts', the pace of scientific research and the demands of new disciplines, had made it 'impossible' for these institutions to keep pace.[81] Though many of the former institutions had existed for only a few centuries, 'at present these old *corps* have ceded their place to others more vivacious, and, one must say today, more popular with the general sympathies'. In describing these old academies, Quetelet quoted the Italian dramatist Vittorio Alfieri who compared the proliferation of gentlemanly academies around European governments to court jesters around medieval princes, except that Alfieri found the jesters 'a lot more useful'.[82] But the nineteenth century, Quetelet argued, had finally seen the creation of practical national organizations, none more important than the Académie royale. Though he claimed the speech was 'no apology' for Belgian academies, he could not hold back his praise of the institution where his national career had begun: 'All impartial men who examine the intellectual history of our country know the effect which has been made on the sciences since the reorganization'.[83] The invigorated Belgian academy shared with its European counterparts the power not only to undertake large-scale research projects, but also to confer legitimacy on scientific claims. After all, Quetelet asked his audience,

> Who is the man who does not accept with confidence new facts once they have received the sanction of the great bodies of *savants*, like the French Institute, the Royal Society of London, the Berlin Academy, etc? Is it not there where the most honour is due, and where one finds the most utility?[84]

Quetelet's praise for the superiority of institutions supported his belief that individual genius was no longer necessary in the sciences. In the same manner, he defended *académie* members for forgoing their own interests in favour of directing large research projects, arguing that no single person could speak with as much authority as the institution itself. In light of his theory on the 'sanctioning' power of academies on the production of knowledge, the words Quetelet used to praise his fellow academy members – '*sans amour-propre*', '*consacré*' and '*absent tous désigné*' – can be seen not just as general praise for hard work, but as a recognition that the researchers and observers who made up the *académie* were the kind of men to lead future scientific projects.

Not all members were equally enthusiastic about the rush to create a new *académie*. It was, after all, an *académie* of *lettres et beaux-arts* too. The relative neglect of the arts and literature after the reformation of the *académie* did not go entirely unnoticed. As far back as 1817, there was concern about uneven membership after a decision to split the institution into two classes: *sciences* and *histoire et littérature*. As the *académie* notes from 1 April 1817 attest, 'one can see that after this division, the number of members attached to each class has not been in proportion to the adopted base. It has been resolved to rectify this situation'.[85] Though this meeting fixed the number of members of the *sciences* at thirty-two and *histoire et littérature* at sixteen, even this unequal distribution did not reflect the disparity in activity. Quetelet's friend Reiffenberg had contributed a few articles on the history of linguistics and philosophy, but aside from the historical essays of the secrétaire perpétuel Dewez, there was almost nothing in the *Annuaire* on literature or the arts. As a later historian of the *académie* wrote, by 1830,

> Certain members saw that the Class of Letters had not attained the scientific level of the Class of Sciences. The members of the latter were younger, formed rigorously by the discipline of observation and were more concerned with lively contacts.[86]

The choice of words here is certainly relevant: the 'discipline of observation' and 'lively contacts' were precisely the two skills that Quetelet wanted to instil in the *académie*. And the inclusion of age as a criterion for 'scientific level' must have surprised a few of the members from the Habsburg days. Conversely, 'rigorous observation' and 'lively contact' were hardly the stock and trade of writers and artists who, like the former secrétaire perpétuel van Hulthem, chose to work their craft in isolation. If the criterion was indeed to attain a 'scientific level' then one might not be surprised that the sciences would be superior to the arts; there is, after all, no record of complaints about the Class of Sciences not attaining the 'literary level' of the Class of Letters. Quetelet, the former poet, now praised members of the academy specifically for a lack of literary flourish. As he wrote of the dedicated observer Cauchy, 'It was not his desire to write in order to distinguish himself, but to render himself useful'.[87] Yet as the 1830 note makes clear, it was the sciences rather than literature that mattered at Quetelet's reformed *académie*.

Based on the concerns of *académie* artists and the increased prominence of Belgian men of science in Europe, Quetelet was successful in reinvigorating certain aspects of scientific life in Belgium. Though he did not become secrétaire perpétuel until 1834, by that point he was already assumed to be the true head of the academy. His nomination, on the day Dewez died, showed that his rise to the top of the *académie* had been pre-ordained for some time. By then the old guard as represented by Van Hulthem had been removed and replaced with a new set of younger dedicated and devoted servants of science. Though cases like Cauchy and Simons – surveyors who had given their life for their work – were extreme, they embodied the general traits for the model men of science Quetelet wanted to produce in Belgium. These men in turn had been able to secure work in government sponsored research, bringing together a fusion of science and politics which approximated the model Quetelet witnessed in France. Just as William and Falck had accomplished their goals with the renovation of the educational system of the southern provinces, Quetelet had by the end of the 1820s changed the entire order of scientific production in Belgium. All that remained was to share the fruits of that reform with the rest of Europe.

Conclusion

Quetelet's experiences at the Observatoire royal de Bruxelles, the Athénée de Bruxelles and the Académie royale reveal several elements of his professional life that defined his later social physics. Primarily, they exhibit Quetelet's remarkable ability to build large networks of researchers out of limited means, or to 'actualize' ideas that had been circulating throughout Europe. In trying to create order and stability in Belgium, he was able to craft messages to both the king and to the general public about the importance of an observatory. He did not invent the idea of an observatory, public science education or a revived academy in Belgium, but he did bring together the people and ideas that would make such things possible. Secondly, these ideas and achievements display his skills in working with governments – perhaps the primary scientific qualification of the day in continental Europe – to meld scientific and political goals. Finally, Quetelet's accomplishments in Belgium help to show the kind of men who could succeed in science in the aftermath of the war of arts and sciences. 'Lively contacts' and a group of young men formed 'rigorously in the discipline of observation' were the order of the day.

Quetelet's work to secure funding for the observatory, to teach astronomy and to reform the *académie* are also evidence that his scientific interests often were driven by his desire to create institutions and that this approach was dependent upon the relatively recent state-sponsorship of science in Belgium. As a 'pure' mathematician in his teens, Quetelet had needed only a pencil and paper as well as a few primary sources in order to conduct his research; the state needed to provide him with nothing more than a suitable allowance for teaching. However, as his interest in astronomy grew, so did his dependence on the state for financial support. In this effort to expand his methods of research, Quetelet had little trouble in convincing state officials to help. So successful did Quetelet become at convincing men like Falck, that it might be wrong in fact to suggest that his 'pure' scientific concerns necessitated his search for state assistance. Rather, Quetelet's ability to convince government administrators led him to pursue scientific projects where he could put his talents of persuasion to good use. Returning to Bourdieu's description of scientific activity from the introduction to this book, Quetelet in Belgium was a master at managing to 'set up the standard most favorable to [his] personal and institutional capacities'.[88] To continue to work in conic sections or geometry, to say nothing of poetry, would have left many of his skills dormant.

Fortunately for Belgian science and industry, Quetelet did not remain a poet or a mathematician. Instead he helped contribute to a host of new technologies and organizations that, directly and indirectly, helped spur Belgium's remarkable industrialization in the nineteenth century. The mines were explored, the railroads built and the canals dug. The transformation in Belgian industry did not necessarily result from a traditional revolution in scientific thinking or even through physical technologies based on recent scientific discoveries, but through organizational changes directed in part by Quetelet. The men Quetelet recruited with *le feu sacré* like Cauchy, Simons and Crahay had done more in a few decades to transform the Belgian countryside than the old men of the *académie* could have accomplished in a lifetime.

4 FROM BRUSSELS TO EUROPE: THE CREATION OF A SCIENTIFIC NETWORK, 1823–9

Beyond the Belgian Frontiers

Throughout the 1820s Quetelet created a set of national institutions in Brussels that would have been unimaginable two generations prior. The decision to focus on creating institutions of science rather than pursuing a career in the arts had been professionally successful, and while the scientific community of the United Kingdom of the Netherlands may not have equalled that found in England, France or the German territories, the reforms of King William I had provided Quetelet the perfect opportunity to explore his interest in large scientific networks. As important as these reforms were, the institutions and practices of scientific researchers across Europe had been changing just as rapidly, with leading men of science like John Herschel and Alexander von Humboldt stressing that science needed to move beyond the state to embrace international collaboration. By 1830 Quetelet may have brought Belgian science to the level of Napoleonic France, but by that time many aspects of scientific research had been internationalized; to truly be part of the new form of scientific practice, Quetelet needed to leave Brussels.

As the dates in the subtitles of both this and the previous chapter indicate, Quetelet did not build up the institutions of science *and then* insert them into an international network of *savants* already formed; rather, the very process of building a national set of scientific institutions depended upon international collaboration and vice versa. Quetelet created the observatory while meeting with the directors of international observatories, and rose in the ranks of the Académie royale at the same time he collected data from colleagues across Europe. This chapter explores then the ways in which Quetelet traded on his own standing in Belgium to expand his scientific network into France and Germany, the two 'frontiers' which had been so instrumental for Henri Pirenne in the nation he called 'half-French, half-German'. Most important for his introduction to the wider world of European science was a formative trip to Paris in 1823 to learn the practice of administering a major observatory, where he met two of the leading administrators of French science, François Arago and Alexis Bouvard. Two years later, Quetelet co-founded the continental journal *Correspondance mathématique et physique*, a publication that provided his first access to a larger set of data on society and brought him into direct contact with demographers like Louis-René Villermé and their recent data. By 1829, when Quetelet embarked on a trip to Germany to visit men of science as diverse as Goethe, Gauss and Humboldt, it was clear that he was meeting with equals. By the end of this period of work and travel, Quetelet, on the eve of the publication of his most enduring work, would be as well connected as any man of science in Europe.

The Able Generals: Arago and Bouvard at the Observatoire de Paris

Quetelet's requests for an observatory may have been laboured, but they resulted in a tremendous opportunity: the chance to travel to the Paris Observatory to see one of the world's model institutions of science. Though King William and the minister Falck may not have initially been aware of the opportunity when they sent Quetelet abroad, France in the 1820s also provided the guide to integrating science and state. The Institut de France (Académie des sciences), the École polytechnique and the Observatoire de Paris were all supported and staffed by a government that had helped make the reputations of most of the great names of French science, including Turgot, Condorcet, Jérôme Lalande, Adrien-Marie Legendre, Antoine Lavoisier, Poisson, Joseph Fourier, George Cuvier and, above all, Laplace. Throughout the tumultuous years of revolutionary government, the Consulate, the Empire and finally the Restoration, the nexus between state control, bureaucratic data collection and the practice of empirical and theoretical science progressed almost unabated. As Alexis de Tocqueville noted after examining the French state in its pre- and post-revolutionary days: 'I thought that the taste for statistics was peculiar to present-day bureaucrats, but I was wrong'.[1] Robert Fox described a similar progression after Napoleon's fall, noting that even as the public face of science changed significantly, one of the 'subsidiary trends' in French science in the nineteenth century was 'the ever-closer integration of science ... within the bureaucracy of the state'.[2] William and Falck had sent Quetelet to Paris with the primary intention to gather information on astronomical instruments for the Observatoire royal de Bruxelles, but his experience with the French model of science proved as formative an educational experience as the technical training.

Quetelet's three months in Paris have been described by one of his biographers as the pivotal moment in the development of his social theory,[3] and it was no doubt true that through an introduction to the probability theories of Poisson, Fourier and Laplace he first learned of the use of the Gaussian 'error curve' in determining the fixed position of astronomical observations. This discovery, when combined with the large amount of population data he would find in Belgian and French records, formed much of the foundation of *Sur l'homme*. Furthermore, Quetelet's praise for Laplace was effusive and Laplace's injunction to 'apply to the political and moral sciences the method of observation upon calculus' served as Quetelet's rhetorical lodestone in the 1830s. As a story of a transfer of ideas from one person to another, the theory of a 'Laplacian Moment' is quite convincing. Yet a reconstruction of Quetelet's time in Paris finds that though he met and admired the probabilists, the vast majority of his recorded experience in the summer of 1823 was spent in the observatory trying to understand the practical demands of running a large institution. In this endeavour, he was far more engaged with the two co-directors of the observatory – Bouvard and Arago – than he was with the great theories of Laplace. Quetelet may have learned about the error curve and theory of large numbers in Paris, but it was his Parisian education in the administration of a government-sponsored scientific institution that most assured his future success.

For one of the most important moments of his career, the twenty-seven-year-old Quetelet had an inauspicious arrival in Paris. Years later he described his nervousness upon first seeing the observatory: 'In entering this illustrious monument of great works, I felt even more aware of what I lacked'.[4] Though he recognized that his principle mission was to learn the administration of an observatory, he also admitted that 'at the same time' he came to Paris 'with the conviction that all of my instruction in the practice of astronomy remain[ed]'. Even walking up to the offices of Arago and Bouvard proved an adventure, as Quetelet, overcoming his fears, 'climbed the great stairs with much assurance', but upon arrival found himself 'between the neighbouring doors of Arago and Bouvard' where he 'stayed for some time indecisive'.[5] After finally deciding to knock on the door of Arago's office, Bouvard emerged from the other office and found a petrified figure in the hallway.

Quetelet had reason to be concerned about his experitse. Despite his modern standing as one of the greatest Belgian astronomers, at the time of his arrival in Paris in 1823 Quetelet had hardly any experience in practical astronomy and in fact had never been inside a major observatory.[6] His experience had been limited to books, reports and journals, which, while satisfactory for the study of conic sections or classical poetry, provided little guidance on how to actually go about manipulating the large instruments of astronomy or organizing a group of assistants. In this first step outside of the provincial science of Belgium, Quetelet recognized just how large a gap existed between the science he knew at home and science as practised in Paris.

Confronted with this challenge, Quetelet responded with his customary enthusiasm and diligence. Rather than being daunted by the great men and institutions of Paris, he was inspired to write to Falck soon after his arrival that 'the formation of a great observatory in our southern provinces appears easier than I had believed it to be'. There was even, he claimed, good reason to establish several smaller observatories throughout the Low Countries.[7] And Bouvard, after presumably recovering from the shock of seeing the young Belgian standing flatfooted in the hall, immediately introduced Quetelet to his friends and offered him lodging. Quetelet quickly became part of Bouvard's important Thursday dinners attended by Laplace, Poisson and the other leading lights of French science, becoming, in his words, 'almost in some sense a member of the family'.[8] On other nights, he would spend hours in the observation room, consulting with the 'numerous foreigners' who shared his passions. Shortly after, Bouvard and Arago pledged to offer support for a Brussels observatory, shoring up any doubts about the future of Belgian science. In just a few weeks after his hesitant arrival, Quetelet judged the mission to Paris a success.

How was Quetelet able to enter into this network of some of the most important scientists in Europe – to convince Bouvard to let him dine with the 'the illustrious author of Mécanique Céleste' – just a few weeks after arriving empty handed and indecisive?[9] First of all, Quetelet's explanation of his state of mind upon arrival in Paris can be questioned: it is difficult to find any external reference to him ever experiencing a lack of confidence. And while it may have been true that Quetelet had no formal letters of recommendation, his advisor at Ghent J. G. Garnier had been Arago's teacher in France years before. Quetelet had also been sent as an official representative of the king, hardly making him an unknown quantity. But even these connections do little to explain why he was so successful in courting the support of Bouvard and Arago. To explain the relationship then, it is helpful to examine the *éloges* Quetelet presented to the Académie royale for both directors. Though much of their affinity came about because of a shared interest in astronomy, Quetelet's notices upon the death of the two men reveal how much the young Belgian learned from his first trip abroad.

The connections to Garnier and King William helped Quetelet gain access to the Thursday dinners, but it was clear that in Alexis Bouvard (1767–1843) Quetelet had found a mentor with common traits. Bouvard would make several important contributions to astronomy, most importantly his initial recognition of Uranus's inconsistent orbit, leading to the shocking discovery of Neptune in 1846. Yet Quetelet's remarks showed preference for the sense of order Bouvard imposed on the observatory. Bouvard's great contribution had come about, in the words of one historian, because of the 'amazing things that can happen when one begins to compile data'.[10] Quetelet even ranked Bouvard near Falck in his hierarchy of 'rare men', calling him '[one] of the type of rare men who have devoted themselves entirely to the sciences and who cultivated them'.[11] Almost thirty years senior to Quetelet and Arago, Bouvard had entered into the great institutions of French science when they were flourishing in the wake of the reforms by Turgot and Condorcet, taking up the directorship of the Paris Observatory in 1793. There he first worked with Laplace, who was using the observatory (and Bouvard) to collect data for *Mécanique céleste*. Quetelet claimed that 'above all', it was this great collection of observations that guided the direction of Bouvard and the observatory. For almost twenty years, Bouvard and Quetelet remained in contact, with the former offering advice and support in Quetelet's long quest to construct his observatory. Like many of his professional relationships, Quetelet also felt a deep affection for the elder Bouvard, claiming that 'he loved me with the tenderness of a father'.[12]

Bouvard appeared a willing servant to science and the state when Quetelet arrived, but it had only come after some convincing. As Quetelet explained, Bouvard was compelled initially by the revolutionary French government to take up the direction of the important observatory, 'with the threat of being imprisoned if he tried to leave'. Despite this unattractive entry and the general 'cruel privations that were imposed upon him', Bouvard turned the Paris Observatory into one of the preeminent scientific institutions in the world, attracting some of the best minds of France.[13] By the time Quetelet had arrived, Bouvard was able to introduce him to an international set of like-minded astronomers and meteorologists, exactly the people Quetelet needed to meet in order to locate his own institution in the international network of observers.

It is an indication of Quetelet's social talents that he was able to work well with such diverse personalities as Bouvard and Arago. Even more impressive given that the two directors hardly spoke to one another due to a disagreement over the election of Bouvard's brother Eugene to the observatory faculty.[14] Though both skilled astronomers, they clashed as personalities, with the younger Arago often frustrated by Bouvard's 'lordly manner'.[15] According to Arago's biographer, the adventurous young writer was initially struck by the dull routines of both the observatory and its director when he first arrived. One of the preeminent scholars of the Paris Observatory, Maurice Daumas, claimed that Bonald:

> lived with all the economy of the *Savoyards*, with an economy of sentiments and space. Only one passion held him: numbers [*le chiffre*]. Bouvard calculated; he lived to calculate and calculated to live. All of himself he devoted to mathematics ... and when he did not calculate, he observed.[16]

Quetelet's father had been from Savoy as well, and the hopeful astronomer certainly appreciated the dedication to calculation and observation. The right man it seemed had emerged when Quetelet had hesitated on the landing outside the two directors' doors.

While Quetelet's praise of Bouvard fits well within the kind of 'new man' he had been trying to create in Belgium, the experiences of the younger and more adventurous François Arago (1786–1853) provided another lesson for the young Belgian. Like Quetelet, Arago had been raised in a pedagogical culture that promoted excellence in both the arts and the sciences and had gradually forgone literary studies in service to the country. His transformation began at the École polytechnique, which under Napoleon was strictly a training ground in practical mathematics and engineering.[17] Though Quetelet had once quoted Arago's friend Berral lamenting the death of the old way of instruction, noting the 'great misfortune' brought about by an 'incomplete education',[18] he was equivocal about the move towards more technical instruction while reflecting on Arago's career:

> I have often discussed the advantages and the inconveniences of the old and the new organizations [of schools]. If the former were more favourable to the development of intellectual faculties, and have given the more illustrious names to the Institute, the latter have perhaps made men more capable for public service.[19]

The sacrifice of 'intellectual faculties' in favour of 'public service' was precisely Quetelet's vision for the future of Belgian science. Many *savants*, Quetelet included, believed in particular that the time of great discovery and exploration of the skies had largely passed in astronomy, and that Copernicus, Kepler, Galileo and Newton had exhausted the study of the great movements of the heavens. While there may have remained a few planets and stars to discover, there was no more need of genius. The national observatory, charged with providing accurate surveys and catalogues for maritime travel and meteorological data, needed dedicated assistants rather than great visionaries. Based on Quetelet's organization of institutions like the *académie*, it seems he leaned towards the 'advantages' of the new schools over the 'inconvenience' of less favourable intellectual faculties.

In this transition from the old 'organizations' to the new form of 'public service' in science, Arago was himself an important and paradoxical figure, whose fame in 1823 was due in equal parts to his achievements in science and adventure. Not only had he been named secretary of the observatory at twenty-two, but he had also lived a far more active life than most of his fellow students, serving time in a Spanish prison, an experience recounted in later years in *Histoire de ma jeunesse* (1854). After escaping his captors, Arago had travelled to Africa as a fugitive and was greeted as a hero when he returned to France. Quetelet even placed him among the 'great French literary tradition' of Buffon, Rousseau and Chateaubriand.[20] Arago's personality also led him to renounce many of the traditional disciplinary honours of French science, including taking up dual membership in the Académie français and the Académie des sciences, or engaging in the lucrative but controversial practice of *le cumul*, where *savants* held several joint academic and state positions (and salaries) to ensure a sustained career.[21] He had long complained of the professionalism and careerism he found in his education at the École polytechnique, and in defending his decision to limit his membership in the academies, quoted a letter from d'Alembert to Lagrange on the perils of institutions.[22]

In his rejection of many of the easier career paths, Arago might seem similar to Quetelet's friends in Ghent and Brussels who complained bitterly of the new forms of scientific research, yet Arago was also an extraordinarily successful administrator. 'In the domain of science', Quetelet later wrote, 'he was like an able general who assigned to each his post, who delivered the order, and indicated the goal toward which one must march'. At times, Arago even went so far as to decline credit for the large land surveys he had directed, telling his employees to put whatever name they wished at the top of the report. Even though Arago had made the crucial suggestion that a planet might be responsible for the perturbations in Uranus's orbit, leading to the discovery of Neptune, he had only sought to promote Le Verrier's claim and not his own.[23] For Quetelet, it was this selfless service to the work and to the state that marked greatness in science. 'These sorts of men', Quetelet claimed of administrative 'generals', 'are infinitely rarer than the special genius'.[24] While Laplace may have been the 'French Newton', his form of genius was apparently not as exceptional as Arago's administrative capabilities.

While Bouvard may have been born to direct the observatory, for Arago the process of transformation from adventurer to professional scientist was much more difficult. Like Quetelet, Arago began his scientific career in abstract mathematics, studying in physics, conic sections and other geometric disciplines that favoured independent study over group data collection. Yet as a student at the École polytechnique, Arago saw immediately that he was being trained for bureaucratic service, a change he initially rejected. Whether it was the culture of the École polytechnique – a feeder school for the Institut de France which essentially determined the career prospects for almost all French *savants* – or the opportunity to work with Laplace or simply a change of heart, he eventually gave up his objections and embraced the practical aspects of mathematical training. As one commenter noted, Arago represented 'the youthful force' ready 'to devote themselves to the joys of mathematization'.[25] Arago's development in the 1820s was also representative of a change of culture in France, as the adventurous days of the Empire gave way to the pragmatic days of the Restoration. This new culture of scientific work was not so much embraced by Arago as tolerated. It was not until the observatory, Daumas noted, that 'the sciences became [for Arago] a career and no longer a secondary occupation.'[26] Quetelet may have recognized in Arago the experience of a transition from a well-rounded education of literature and abstract mathematics to a more practical course of education. Indeed, the École polytechnique, in its service in developing useful *savants* to the state, became the model organization for Belgian technical education in the same way the Paris Observatory was the model for the observatory in Brussels.

Quetelet continued to admire his friend's abilities even after leaving Paris, when Arago's administration of the Académie des sciences came under attack. Like Quetelet's friend the Baron de Reiffenberg, Arago had been criticized for taking on too many responsibilities across too many different disciplines. Yet the criticisms reflected what Quetelet appreciated most about Arago's ambition. In defending his friend against charges that Arago spent too much time on administrative duties, Quetelet expressed some of his strongest sentiments on the need for powerful officials among the ranks of government scientific institutions. He began by claiming that 'Arago was certainly one of the most honest and disinterested men who have ever been employed in public service'.[27] Contesting the claim that Arago spent too much time in official positions, Quetelet wrote that 'these accusations are often made against men who make all their fortune in scientific progress and who make infinitely more considerable returns than those *savants* who take for their constant goal to attack them'.[28] Quetelet even protested that service to the state could equal the intellectual work of science; that Arago was not 'wasting his time' in government:

> Arago can be rebuked for the part which he took in public affairs; one can, in effect, regret to see a *savant* of such merit waste his time on things that are of secondary order. But in doing so, he is in a position to spread useful light on important questions, which, were they to be resolved by incompetent men, would compromise the future of the county.[29]

Even if there was a loss for science because *savants* like Arago took up public affairs, Quetelet countered that a man of genius could do more for science by 'speaking out' publicly and helping it 'to receive the support to which is has right'.[30] By defending Arago, Quetelet also provided the perfect defence for his own scientific interests in Belgium.

In light of Quetelet's comments on Arago and Bouvard, it seems that Quetelet learned more from the Paris Observatory than how to reduce a set of observations or to calculate the longitudinal distance from Brussels to Paris. He also learned the process of administering an observatory. While Bouvard may have been a father figure to Quetelet, confirming in the young Belgian the benefits of careful and patient observation and calculation, Arago had showed him the levels of power a well-connected *savant* could reach, even one who had begun his career by rejecting a narrow education. Quetelet never reached quite the political heights of his friend Arago, the future prime minister of France, but he was able to learn the value of administration in science.

This reading of Quetelet's trip to Paris should not be meant to dismiss the ideas on probability, observational astronomy and the application of statistical models to society that Quetelet encountered in his Thursday night dinners; his time in Paris did prompt Quetelet to investigate social phenomena with astronomical and statistical methodology. However, among the important things Quetelet took from his experience in France, an appreciation of the error curve is not apparent from his own reflections. Primarily he learned about the equipment needed for a large observatory and the administrative details of operating a large institution.

The trip to Paris also taught Quetelet the power of connections in nineteenth-century science. His advisor Garnier led him to Bouvard, and Bouvard to Laplace. In just two degrees of separation, Quetelet had managed to move from the limited world of his quaint medieval hometown to the company of the most famous living *savant* in Europe. From the perspective of Quetelet's theoretical *l'homme moyen* and *physique sociale*, Paris was the place where he encountered the idea of applying probability theory to human actions. But, from the perspective of the time, Paris was much more important to his education in the developing practice of science in Europe. Arago and Bouvard, with whom he would keep up a steady correspondence, were notable mostly for their expertise in operating an observatory, not for theoretical advances. Paris was a formidable trip for the young Belgian, but not as much for what it taught Quetelet about new scientific theory, as for what it reinforced in his own beliefs about the importance of administering science.

Correspondance Mathématique et Physique

If the 1823 trip to Paris convinced Quetelet he was ready to direct a large network of scientific observers, he still faced the problem of having no institution from which to direct these investigations. The observatory in Brussels seemed permanently on hold, the *académie* membership was too diffuse to be completely controlled by one person and Brussels offered no other position that allowed Quetelet to collect information and maintain contact with researchers across Europe. In 1825 however an opportunity to direct a continent-wide publication appeared when his former dissertation advisor Garnier proposed they co-edit a new journal. It might not have offered the same opportunities as a modern observatory, but it would have to suffice in the interim.

Second only to his position as director of the observatory, editing the *Correspondance mathématique et physique* was the primary means through which Quetelet acquired his formidable reputation outside of Belgium. While the journal began as a clearinghouse for recent publications by professors at the six universities of the Low Countries, by the end of Quetelet's tenure it attracted notices from noted French demographers like Lois-René Villermé and members of the British Royal Society like Charles Babbage and Edward Sabine. Consistent with Quetelet's career, the expansion of his network of associates was coordinated with an expansion of his interests. Indeed, the *Correspondance* contains the earliest drafts of many articles that would later make up *Sur l'homme*, and in Chapter Five specific attention will be paid in to the statistical and demographical works that appeared during this time. For the purposes of this section, however, the content of the *Correspondance* reveals the scope of Quetelet's activity during the time he awaited the finished observatory. Though comparatively little attention has been paid to this section of Quetelet's career, his fourteen years as editor allowed him to keep contact with his associates from Paris and London and also provide the foundational evidence on which he would build his *physique sociale*.

The first volume of the *Correspondance* in 1825 offered few hints of what was to come, with almost all the subscribers coming from the university class or members of the Académie royale. All but one came from Belgium. The interior minister Van Ewyck subscribed, as did Quetelet's old friends from Ghent, but the initial run was small, especially given the relatively inexpensive subscription price (seven florins) and the important gap it was meant to fill in Belgian intellectual life. As Quetelet and Garnier wrote in the prospectus, Belgium had plenty of medical and natural science journals but no national publication dedicated to research in the abstract and applied physical sciences.[31] Though in 1825 Belgium could hardly be said to be overflowing with important research, the lesson that Quetelet had learned in Paris, and one he had used in making the case for the observatory, was that scientific knowledge followed from the creation of institutions and networks of science, not the other way around.

Though designed ideally as an outlet for publishing new ideas, the *Correspondance* concurrently served to educate Belgian *savants* in the sciences; not only in recent research, but also in the fundamental theories of mathematics and physics. Initially divided into two sections of *mathématique* and *physique*, each 'field' also contained a subdivision between 'reviews of scientific work' and 'questions', the latter of which to the modern eye resembles problem sets.[32] Quetelet wrote the reviews in sections on 'Astronomie', 'Géométrique Analytique' and 'Physique Mathématique' and Garnier oversaw 'Mathématique Elémentaire', which asked readers to solve algebraic equations and send the journal their solutions. Each quarter, Garnier published the best solutions along with commentary and, in turn, professors wrote in to correct Garnier. Though original research from the provinces was published, the *Correspondance* at the outset resembled a textbook as much as a peer-reviewed journal.

At the same time Garnier was working to educate Belgian scientists, Quetelet was using the journal to strengthen his ties to other scientists in Europe. He reprinted letters from his friends from Paris, Arago and Bouvard, including a mention from Bouvard of the new *lunette parallactique* installed in Paris, the same expensive technology that Quetelet had hoped to acquire for Brussels.[33] But, oddly, Quetelet did not write the majority of the notices on French work in probability and statistics. The first issue included three pieces from France that would in retrospect have seemed to fall under his natural domain: a review of the fifth edition of Laplace's *Système du monde*, a short notice by Villermé on the striking effects of poverty on mortality and a commentary on Laplace's *Essai philosophique sur les probabilités*.[34] Yet the last two notices were written by Garnier. Villermé's population studies and Laplace's probability theory played an important role in Quetelet's later social physics, but in 1825 his focus was still on the observatory and his network of fellow directors.

In the next few years of the journal, Quetelet's interest in statistics broadened as the two editors began to cede more of the writing to foreign correspondents. As *savants* from across Europe began to contribute and Garnier left the journal in 1827, the *Correspondance* became Quetelet's most important educational experience in overseeing a large set of observers. Among the most notable contributors was Villermé, who sent the journal the wealth of data he had accumulated from Parisian criminal records, including data that would later feature in the efforts to combat the 1832 cholera outbreak in Paris.[35] In the second volume, the French demographer wrote to congratulate the editors on the inclusion of a *Statistique* section and to share with Quetelet his novel idea that Lent affected criminal behaviour. Years before Quetelet would make any such claims, and long before the popularization of polling as a guide to the public will, Villermé wrote that 'the morals of people and the measure of its opinions, are therefore written sometimes in the results of Statistics; it is only necessary to know how to read them!'[36] A year later, Jacques de Mainville wrote to claim that 'it should be seen, M. Quetelet, that statistical work can be used not only on men's physical behaviour but on their moral behaviour as well'.[37] Though it is true that the idea of extending the methodologies and practical applications of the natural sciences to moral behaviour had been proposed since at least Condorcet and Laplace, no one until the early nineteenth century had accumulated sufficient data on populations to actually conduct this research. Villermé and Mainville provided Quetelet with no new conceptual advances, but they did provide him with data, something he cared far more about than a new scientific theory and something he desperately needed while the observatory project floundered.

In his writings on statistics between 1825 and 1829, Quetelet complained often about the paucity of data for statistical investigations of society. Conversely, what is sometimes taken to be his most important idea from the time – applying probability to social and administrative statistics – received little attention. This was not because Quetelet did not believe in the expansion of mathematical approaches towards social phenomena, but rather because such a programme was so immersed in European ideas at the time as to seem unremarkable.[38] Instead, the major challenge to Quetelet's project to construct a statistical account of social behaviour was not intellectual, but practical – he needed numbers. As he wrote in 1827 in a work that drew together his research for the *Correspondance* and which consisted mostly of tables, he had little to add to the limited raw data, hoping only that it '[could] inspire authors who have the taste and necessary patience to provide more traction'.[39] Though Quetelet had great hopes for a mathematical science of social behaviour, at the time his plans were much less ambitious. He would be 'pleased if this new research ... engag[ed] friends of the sciences to do more'. This alone', he wrote, 'can assure it a useful character'.[40]

The 'friends of science' responded to his call. Over the next few years at the *Correspondance*, he received a staggering amount of data on birth and mortality rates, criminal prosecutions and other administrative statistics. Some came about due to the 'avalanche of numbers' created by statistical associations across Europe in the decades after the Congress of Vienna, yet much of it came from Belgian and Dutch bureaucrats or others whom Quetelet knew personally.[41] When Quetelet began in 1825, the only numbers had come from a mortality table put together by Willem Kerseboom in Holland for the purposes of determining life insurance.[42] He had published a short piece in the first volume of the *Correspondance*, but without comparable data, there was little to say other than to give a simple descriptive account. Initially frustrated, Quetelet used his contacts from the *académie* to try to find data in Brussels, asking his student and fellow academician Pierre-François Verhulst to put together a similar table for the Belgian capital. More importantly, Quetelet also included a small notice on statistics in the first volume from Francis Baily. Baily had published his work in the *Annuaire* of the French Académie des sciences in February of 1825, but by the beginning of 1826, Quetelet had still only read an abstract.[43] At the least however it proved that there were other European researchers working on the problem.[44]

Quetelet had created the statistics section with little in the way of data but again the information followed the institution. The year 1826 alone brought in a number of outstanding reports that drew from the established bureaucracies in the United Kingdom of the Netherlands. From a former friend and classmate from Ghent, Jean-François Lemaire, Quetelet now had access to mortality rates in the Belgian city of Tournai for the past nineteen years. Lemaire in fact owed his success to Quetelet's career and, like many of Quetelet's friends, had forsaken a life as a writer in order to take up a position in King William's administration.[45] Comparing Lemaire's data with a notice from Villermé in Paris allowed Quetelet to find regularity across geographically distinct areas.[46] In the north the bureaucrat and *Correspondance* subscriber Reheul Lobatto provided an even more extensive data set, the sprawling and dense *Jaarboekje*, which had been financed by William and included all manner of records, from meteorological and astronomical data to marriage records. Though scattered and unorganized, Quetelet suggested the *Jaarboekje* as a model to be taken up by correspondents across Belgium and Europe.[47] Finally that year, Quetelet reviewed the *Description statistique de la Guelde*, a work that combined data on 'meteorology' and 'commerce' with the 'social and religious habits' of the inhabitants of the Dutch province on the Rhine. Though both the *Jaarboekje* and Guelde data included almost no analysis, Quetelet concluded confidently that these figures could 'tend to give an intimate knowledge of the physical and moral state of the province'.[48]

Quetelet's faith that raw data could lead to profound conclusions about the 'moral state' of populations might seem strange, especially given the fact that almost all of the tables appearing in the *Correspondance* came without any analysis or conclusion. Yet his guiding principle in science was always to build the institution or network first and then collect the data. As seen in Chapter Five, the theories would eventually emerge out of this data, but the *Correspondance* gave few hints of what was to come. What is surprising from these early reports to the *Correspondance*, at least to those familiar with Quetelet's later work on *l'homme moyen*, was how little the data had to do with human actions and how indistinguishable population data was from physical or meteorological data. While there were separate sections for 'Meteorology' and 'Statistics' it was often hard to know which one was which, as many of Quetelet's correspondents sent in reports that went back and forth between the weather and the 'passions' of man. As he learned from his friend Lemaire, reports going back to 1781 had tried to show 'the effects of the influence of *météores atmosphériques* on births and deaths'.[49] Quetelet also learned in these years of Moreau de Jonnés's work on the deforestation of English land and its intriguing relation to an increase in temperature and destruction.[50] Finally, Quetelet's reliable reporter Villermé noted there was a clear relationship between the 'march of the seasons and climate' and 'the intensity of fertility'.[51] Looking back at this data through the lens of *Sur l'homme*, or from the perspective of modern enthusiasm for quantifying human behaviour, it might be possible to project some larger plan in mind, but there was no indication to any reader of the *Correspondance* that one set of statistics was more rigorous than another.

When Quetelet took over as the sole editor in 1827, his intentions became clear. Gone were many of the problem sets and more abstract reports Garnier had supervised, and the journal began to publish far more tables and accounts of sciences that netted reliable statistics. One example came in the 'Astronomy' section, a field far more methodologically advanced than population statistics or meteorology, and one where, presumably, mathematical and theoretical advances were taking place. Yet it is striking that so little space was devoted after 1827 to new research in astronomy. The minister Van Ewyck, a subscriber who had been sceptical of Quetelet's scientific justifications for an observatory, would not have been persuaded of possible discoveries had he read the astronomy section of *Correspondance* from the period between 1825 and 1829.

In the first year of the journal, Quetelet included research on shooting stars but also reprinted his letter to William on the national importance of an observatory.[52] After the official order had been signed on 8 June 1826, he wrote in the *Correspondance* to praise William, a leader who could be judged favourably 'by the number of useful monuments he founded'.[53] Of the six articles in the astronomy section published the next year, five concerned descriptions of observatories or instruments, including a long piece on German observatories. This last article involved no research produced by the observatories but only descriptive accounts of the rooms and equipment of four buildings, including such details as where the assistants slept and where the servants' quarters were located.[54] In the same year, in an optimistic spirit, he listed the equipment he hoped to purchase for the Brussels observatory (see Figure 4.1): a *cercle mural* from the 'celebrated English artist' Troughton like that found in the famous Royal Greenwich Observatory; a *lunette meridian* made by 'the premier artist of France' Gambey which had been exhibited in the Louvre and was used in the Paris Observatory; and an *objectif*, *equatorial* and *telescope réflecteur* from the Haarlem Exposition. All of these instruments, Quetelet wrote, would 'give a new proof of the generous protection which had been accorded to the sciences' by William.[55]

MATHÉMATIQUES APPLIQUÉES.

ASTRONOMIE.

Sur plusieurs instrumens destinés à l'observatoire de Bruxelles.

Pendant qu'on poursuit avec activité la construction de l'observatoire de Bruxelles, S. M. le Roi des Pays-Bas a voulu donner une nouvelle preuve de la protection généreuse qu'elle accorde aux sciences, en ordonnant l'acquisition de trois grands instrumens astronomiques destinés à cet établissement. Le célèbre artiste anglais *Troughton*, est chargé de la confection d'un cercle mural de six pieds de diamètre, semblable à celui qu'il a construit pour l'observatoire royal de Greenwich : il est chargé de plus de construire un équatorial de grande dimension, semblable à celui avec lequel *M. South* a fait ses belles observations des étoiles doubles et multiples; et dont on trouve une description dans les *Transactions philosophiques* de la société royale de Londres.

La construction de la lunette méridienne est confiée aux soins de *M. Gambey*, qu'on peut regarder comme le premier artiste de France. Elle sera semblable à celle qui vient d'être exposée au Louvre, et qui est destinée à l'observatoire royal de Paris. La lunette de cet instrument sera de 7 pieds 4 pouces de longueur, et son ouverture de 6 pouces : le corps de la lunette est composé de deux tuyaux coniques dont les bases ont douze

Figure 4.1: Astronomy section of the Correspondance. Though listed under 'applied mathematics', the astronomy section of the Correspondance rarely included new theories or discoveries. More often, as is the case here, the section contained descriptions of observatory equipment or plans to build new observatories. Correspondance mathématique et physique, 3 (1827). Courtesy of Boston Public Library Rare Books Room.

By 1829 the astronomy section contained just two articles (compared to ten in 'Statistics'), neither devoted to new research. This from a journal edited by a man who was director of an observatory and today described primarily as an astronomer in standard histories of science. The three months Quetelet spent away from Belgium that year may partly explain the lack of information on astronomical research, but there was still substantial publication of original material on statistics and meteorology at the same time. Both articles from 1829 were in fact *descriptions* of observatories rather than research reports: one, a part of a longer series on Quetelet's trip to the great observatories of England, and the second, a reprint of a report on the need to build a new observatory of Geneva. This latter article, written by a M. Puerari, began with a question that was of great interest both to Quetelet and his sponsors: 'What is at the heart of the wish to possess an observatory?'[56] This had been a question Quetelet had been trying to answer in his own country for at least six years, and it is not difficult to imagine Puerari's argument as a proxy for his own. As Quetelet wrote in the introduction to Puerari's piece:

> We think that readers of the *Correspondance* will see with pleasure some passages of this writing [to be] dignified ... by the highest reputation that Geneva has acquired for itself. Fortunate are the cities administered by magistrates so clear-headed, and with much reason they appreciate the influence which the sciences exert on the well-being of citizens, and the relief that they give to the country which nobly receives them.[57]

The reader of the *Correspondance* might have assumed that Belgium would benefit from these same 'clear-headed magistrates'.

While the astronomy section languished, the statistics articles kept up a strong pace in the late 1820s. Though publishing had been interrupted by the revolution (there were no editions in 1830 or 1831), by the early 1830s Quetelet had commissioned, received and (less frequently) collected almost all of the statistical research that made up the majority of *Sur l'homme*. Reading the *Correspondance* and *Sur l'homme* makes it difficult to see how Quetelet could have had many new coherent conceptual ideas in the early 1830s; much of what appeared in 1835 was verbatim from his writing in the *Correspondance*. It would be incorrect to view the *Correspondance* as the means through which Quetelet achieved the end of some specific research programme, however, unless the programme can be construed so broadly as a simple call for more data. Instead, Quetelet's later 'discoveries' of social laws appear to have been an artefact of the two interests that would dominate his career and which were cultivated during his time at the journal: engaging with European men of science in exchanges of large amounts of data and educating Belgians on the progress of scientific research around them. As he would write often in his early work on social statistics, the field was so undefined that one could only acquire the data and hope. As far back as the first analysis of the Kerseboom data, for example, Quetelet admitted that he could do no more than 'be struck by the regularity' of mortality.[58] 'Regularity', provided by the data of his friends and connections across Europe, was apparently enough.

A 'Country Curious in Many Regards': Goethe, Gauss, Humboldt and other Rare Men of Germany

By the end of the 1820s, Quetelet had achieved remarkable progress towards his goals: the observatory was on the path to completion, the Académie royale had been reformed and the *Correspondance* had created an international network of researchers willing to assist Quetelet in collecting demographical and meteorological data. By end of the decade, when he departed for the German states to visit their many new observatories, he was far from the self-described timid student of 1823. Unlike his first trip to the observatory which found the young Belgian in awe of the great minds of Paris, the exchanges on this second trip were between equals. The observatory directors and astronomers Quetelet met in Hamburg, Berlin and Leipzig were no longer mentors but colleagues, as interested in meeting '*le savant belge*' as Quetelet was to meet Bouvard and Arago.

Quetelet's travels throughout the German territories might even have offered him an interesting point of reflection on his past, present and future projects in 1829. In this trip he spent an afternoon in the company of a literary hero, calibrated measurements with the great mathematician whose error curve would inspire his most famous theory and lunched with the leading exemplar of international science. In addition to meeting Johann Wolfgang von Goethe, Carl Friederich Gauss and Alexander von Humboldt, Quetelet encountered more of those seemingly ubiquitous 'rare men': observatory directors who dedicated their lives to the supervision of observation. In the wake of his revitalization of Belgian science and his entry into international science through his trip to Paris and the *Correspondance*, Quetelet's German trip was less an exploration of new ideas than a confirmation of the possibility of his most ambitious dreams.

He left from Brussels on 11 July 1829 with a long and professional itinerary planned. 'My principle goal in making a scientific journey to Germany', he wrote, was 'to visit the most remarkable observatories and familiarize myself with the state of astronomy in a country curious in many regards'.[59] A simple enough task, though, like Paris, the practical concerns of learning the techniques of professional observation entailed a more thorough immersion in the culture of scientific activity at the time. Though the region did remain in some ways a mystery – de Staël's 'discovery' of the country's literature had been published in Belgium only fifteen years prior – Germany had been integrated into the network of astronomers for half a century. Quetelet, who spoke excellent German and had this time managed to procure proper letters of introduction, had little trouble navigating the scientific and administrative culture of the 'curious country'.

His adjustment was made easier by his first stop in Altona, a small Danish town on the Elbe that bordered Hamburg. Raised high above the river, Altona had attracted German astronomers since its construction in 1823 and provided a better example of the possibilities of incorporating new equipment into modern architecture than the relatively ancient Parisian observatory. If Quetelet entertained any doubts about the journey, they were soon put to rest upon his first introduction. The observatory director in Altona, Heinrich Schumacher (1780–1850), had like Quetelet began his career as a mathematician but had been drawn to observational astronomy because 'he wished to join theoretical knowledge with that of practice'.[60] Schumacher not only had 'a veritable passion for instruments', but had also succeeded in convincing Frederick VI to provide him with large subsidies to continue his work. Schumacher also assisted Quetelet through providing him introductions to Heinrich Olbers and his former mentor Friedrich Gauss. Just a few days into his travels and Quetelet had already met a well-connected bureaucrat whose greatest strengths were securing government support and observatory practice. It was going to be a good trip.

Though Schumacher helped Quetelet learn some of the new technologies of observation, he also possessed talents of observation in the mould of Bouvard. In his two accounts of Schumacher, Quetelet offered his readers tales of the German's 'magnificent collection of *théodolites*' but also extolled his success in maintaining state sponsorship. Schumacher's father had been an ambassador to the Danish King Christian VII, and his son had carried on the tradition of royal service. In describing Schumacher, Quetelet paid as much attention to personality as he did to ability: 'M. Schumacher functions in his government with a confidence that is justified as much by his work as by the nobility of his character'.[61] Indeed, when the death of Frederick in 1839 had put the future financing of the observatory into question, Schumacher wrote to Quetelet to assure him that he was in good standing with the new king: 'The king has given me public proof that he regards me with the same favour as his predecessor'.[62] Quetelet, who by that time had finally succeeded in courting his own royal favour, knew Schumacher's talents well. As he wrote, 'Schumacher worked with great favour next to the king. To do this well, one must always act with dignity, either to obtain distinctions in favour of *savants*, or to earn accolades for work useful to science'. So great was Schumacher's success that Quetelet referred to his abilities as an art, writing that 'this art characterizes a superior mind, and becomes in the State an element of order for emulation'.[63]

Though Quetelet planned on continuing his travels alone, he impressed Schumacher enough that the German astronomer agreed to accompany him to Bremen in order to introduce Quetelet to the celebrated astronomer Heinrich Olbers (1758–1840). Olbers was at the time one of the most famous astronomers of Europe, having recently discovered what many had thought to be two additional planets of the solar system. Though Pallas and Vesta turned out to be asteroids, Olbers's discoveries were a major boost to the astronomical profession, which had been searching to justify itself as a practical science able to make new discoveries and not just the methodological model for other, more productive, sciences.

Though in awe of the discoveries themselves – Quetelet reports that that he asked 'to touch' the *pendule* that had been used for the planetary discoveries – Quetelet spent relatively little time in his account on Olbers's many discoveries. The two would later team up to try to solve the mystery of shooting stars, but Quetelet had little to say about the man outside of his work to run the observatory.[64]

Perhaps Quetelet's brief account of Olbers was due to a lack of interest in the discoveries of astronomy, but it is also possible that Quetelet was eager to move on to his next stop: Berlin. In the Prussian capital Quetelet found a city unlike any other in Europe in its dedication to institutions of scientific production and education. Even the lack of a suitable observatory – the one in the city dated to 1711 – did not bother him. As he remarked upon his arrival on 28 July, 'few cities offer such a brilliant gathering of educated men in all the different branches of human affairs'.[65] Listing seemingly every institution in the city ('the Institute of the Deaf and Mute, Institute of the Blind, institutes of surgery, medicine, etc.'), Quetelet remarked on the 'particular merit' of Berlin: its focus on the sciences of 'design' and 'mechanical arts'. In other words, he located the strength of Berlin's educational system in its dedication to practical technology. So strong was the Prussian interest in practical education that Quetelet even offered a slight criticism: 'one can complain that letters, geography, and music are too neglected', perhaps remembering the complaints of de Staël from his youth. Quetelet was certainly right that the arts were neglected in the city, but this did not change his overall appreciation of Berlin. Just a few sentences after his remark on the lack of arts instruction, he summed up his position on the city: 'it can be easily appreciated that this beautiful city [is] called upon to take a brilliant place in the world of knowledge'.[66]

Lacking a modern observatory, Quetelet made due with meeting officials from two areas that had recently taken up his attention, the prison system and the Statistical Committee of Berlin. He was particularly impressed with the director of the latter institution, Johann Gottfried Hoffman, who was dedicated to administrative statistics. Though German statistics at the time were primarily descriptive and involved very little application of the ideas of probability from France, at this point Quetelet was more interested in the acquisition of data than in its mathematical manipulation. Hoffman, who 'principally occupied himself with territorial division, population, birth-rates, marriages, and mortality', had access to data Quetelet could only dream of in Belgium.[67] Not only was there an annual printing of population data in the Berlin newspapers, but Hoffman's department received generous support from the Prussian government. 'The king of Prussia testifies by his acts a true desire to spread light in his states', Quetelet wrote, praising Frederick's 'considerable sums to the acquisition of objects of art and science'.[68] The Prussian capital may have lacked a suitable observatory, but Quetelet still found much to admire in the heart of an administrative revolution.

Quetelet was fortunate to arrive not just during the height of Berlin's educational and administrative reforms, but also during one of the rare times that the city's leading scientific star, Alexander von Humboldt (1769–1859), was in residence. Though time was short, Quetelet reported that the two spent an enjoyable afternoon discussing natural history in the garden of the composer Felix Mendelssohn.⁶⁹ Perhaps the most famous man of science in Germany, the globe-trotting Humboldt may at first glance appear to be the opposite of the science workers that Quetelet had promoted in Belgium. Yet Humboldt, who could not have been more unlike the 'calculating' Bouvard or the arch-administrator Falck, possessed the quality of leadership that was crucial to the organization of large research projects. As Quetelet had recognized, the demands of science went beyond what any one individual could do. Humboldt, as Quetelet insisted in a later *éloge*, was yet another example of the new form of scientific activity.

> A geometer can, if he wished, reconstruct all of geometry without recourse to foreign assistance. On the contrary, for other knowledge it is not the same. An ensemble of men is necessary, and these men must adhere to the same voice; it is in some sense the voice of a general who commands: everything depends on the success of these orders.⁷⁰

Like the 'able general' Arago, Humboldt was capable of leading the 'ensemble of men' brought together through international organization and institutions. Arago had in fact been a student of the Prussian explorer and geographer, and Quetelet considered Humboldt to be another 'rare man' of science. Like the leaders of observatories, Humboldt's talents were 'rare, because independent of natural talent and understanding, they must have an agreeable mind which can rally all the ranks and know how to inspire the confidence and desire to lead'.⁷¹

One reason why such generals were necessary was that much of the work of early nineteenth-century observation was, even to the most passionate observers, tedious business, with both observers and computers complaining about the long hours and 'boring work'.⁷² While there may have been people like Arago, Bouvard and Quetelet who enjoyed spending long hours recording data on wind speed, barometric pressure and the position of stars, the battle to gain knowledge in astronomy and meteorology needed soldiers as well as generals. And the obvious place to look for soldiers was the state. The man of science, Quetelet wrote, 'must look for support in the governmental administration [of Germany] … if [they] desire to have an idea of the protection that a country gives to the sciences, letters and arts'.⁷³ Humboldt, a true international man of science, had benefitted from both government support and a collection of helpers.

Not all German science was quite this exciting. After leaving Berlin, Quetelet was disappointed with what he found in the next leg of his journey through Saxony, Dresden and Leipzig. Though the last had one of the most modern observatories, including three items Quetelet would acquire for Brussels – 'a quadrant from Rotterdam', a Troughton Company circle and a new pendulum – his overall tone was that of frustration. In Saxony, Quetelet found a great deal of astronomical and meteorological data, but none that had been observed at the same times as those in Brussels, making comparison useless. 'In seeing so much work made with such laudable perseverance', Quetelet lamented, 'one cannot stop themselves from regretting that science has not until the present drawn more profit'.[74] Hard work and data collection was good, but standardization was still needed.

His spirits would rally significantly, however, due to a fortunate diversion on the way to visiting the University of Halle. Quetelet arrived in Weimer on 27 August to a surprise: the day before the city had held a massive celebration for 'the immortal author of Mephistopheles'. A planned overnight trip had turned into an eight-day stay in which Quetelet had the opportunity not only to meet his literary hero, but also to engage in long discussions in the poet's garden on all topics of arts and sciences. Though Johann Wolfgang von Goethe (1749–1832) was 'perhaps the greatest poet we have' according to Quetelet, he was more interested in discussing the latest scientific advances than the arts. 'He expressed a desire to see the equipment with which I had observed terrestrial magnetism', Quetelet claimed, and the two exchanged ideas on Goethe's recent work on the theory of optics, which he had hoped to get published. Conversely, when Quetelet tried to discuss the Romantic poetry of his youth, he reported that Goethe 'was not too excited', and quickly tried to turn the discussion back to Quetelet's current research.[75] On this latter score, Quetelet clearly must have impressed the amateur scientist. Not only did this meeting spark an active correspondence over the next few years, but after just a few days, Goethe invited Quetelet to be his personal guest at a performance of *Faust*.

As well as being one of the great writers in the history of the German language, Goethe was also one of the most prominent amateur men of science, working outside any institutional home and as much inspired by the ideas of the Enlightenment as he was an inspiration to the Romantics. Though he had attracted a few disciples, his scientific investigations had been ignored at best and ridiculed at worst. In particular, he was roundly mocked for many years because his anti-Newtonian theory of colour was an 'incredible blunder'.[76] His ideas for science, drawn from *Naturphilosophie*, however, were in some respects amenable to Quetelet's programme for quantitative approaches to studying society, emphasizing time, patience and, above all, data rather than the few controlled experiments favoured by Newtonians.[77] Yet if Quetelet recognized any affinities he had with Goethe, he failed to comment on them. After all, Goethe may have represented the height of Quetelet's youthful literary ambitions, but he was close to a punch line in the nineteenth-century science networks Quetelet hoped to join. As Goethe explained to Quetelet: 'As a poet, my path has been made, I can travel it with assurance; but as a *physician*, it is not the same, and the opinions have often varied on my research'.[78] This last point was putting it mildly, but the two still had enough in common to maintain an active and engaging exchange of letters for many years.[79]

Quetelet's exceedingly broad range of interests and social talents can be confirmed by the equal enthusiasm which Quetelet showed for his next host, the mathematician Carl Friedrich Gauss (1777–1855). Though the Romantic Goethe had been Quetelet's idol, by the end of the 1820s, he shared much more in common with the sober and conservative mathematician of Göttingen. According to one biographer, Gauss found Goethe's 'mode of thought' unappealing, and dismissed most all of Romantic poetry, including Quetelet's other hero Schiller, whose 'philosophical views were totally repugnant'.[80] Yet varying judgements on literary ability were not on the table for the two men, as they quickly bonded over a discussion of magnetism. The day after he arrived, Quetelet and Gauss met in the observatory garden, each taking measurements of the earth's magnetism. Though they had slightly different means of calculation, to their mutual delight, the measurements turned out to be nearly identical. As Gauss remarked, 'these observations contain the precision of astronomical observations'.[81] Writing later his friend Olbers, Gauss was impressed with Quetelet and his 'splendid apparatus' which allowed for near unanimous readings of 'the intensity of the magnetic force'.[82]

Magnetism eventually proved helpful in establishing the observatory and provided one of Quetelet's great projects in his later life, yet Gauss's 'normal' approximation of error from astronomical observations – the development so important for Quetelet's early social statistics – was not commented on, neither in Quetelet's remarks on Germany nor in his later éloge on Gauss.[83] Quetelet claimed that far from being moved towards something of a probabilistic turn, his 'travels to Göttingen' carried him 'less towards astronomy ... than towards research on magnetism'.[84] Though Quetelet did apply the Gaussian curve to social data, it is hard to see this as a clear application of astronomical methodology to social phenomena. Similar to Quetelet's experience in France, in which he showed more interest in the administration of science than in the theories of Laplace, Quetelet was drawn in Göttingen to the research field that allowed the greatest opportunity for collaboration. The error curve may have later proved useful for population statistics, but terrestrial magnetism and other earth sciences would provide an even larger potential for data collection.[85]

After meeting three of the most illustrious names in the history of German art and science in the span of just some sixty days, one might think that Quetelet's final passage though Cassel, Frankfurt and Heidelberg would have been anti-climactic. After all, he had learned from some of the great astronomers of the country, seen the city with the most helpful government on the continent for administering science and swapped magnetic measurements with the 'German Archimedes'. Yet there was a final stop on Quetelet's tour, one that for the great scientific organizer was a fitting conclusion to his journey. After almost a decade of traveling through Europe to meet astronomers and mathematicians, Quetelet saw that there might be an easier approach to meeting international men of science. It was called a 'conference'.

The meeting of naturalists in Heidelberg may not have been the first scientific conference, but Quetelet's reaction upon arrival certainly indicated its novelty.

> If I am permitted to make a trivial comparison, I would say that these sorts of meetings are like bazaars, where each carries the product of their work which they love and receive in exchange that which has been made by others. These communications are made more rapidly and in a manner surer than by journals, which often only contain inexact and insufficient details.[86]

Quetelet may have had very little to offer to a conference of this sort, but he took great care in noting the details, specifically the number of meetings, general lectures and 'outings'. Particularly attractive was this last activity, where attendees went for long walks in the woods around the city or took guided visits to the scientific academy. Quetelet had no formal role at the conference, but he did have one responsibility: reporting back to Goethe on the reception to his work. Quetelet knew Goethe was concerned about his reputation as a scientist, and the poet had told Quetelet, 'I am a friend, and I confess that I will be very curious to know what one thinks of these goods [*merchandise*] and if you give it any esteem ... Promise me that you will tell me the truth'.[87] Unsurprisingly, in both his account of Goethe and the German trip, Quetelet reported little on the scientific content of the meetings. Aside from describing the form of the meetings themselves and Goethe's hopes, his only other comment was to mention that he was able to meet the Englishmen Robert Brown and William Whewell, two more prominent men of science with whom he began a lasting correspondence.

This is not to say that the conference was unimportant however, as in later writings Quetelet extolled the virtues of the *congrès*. In the same 1852 essay where he had lauded academies, Quetelet explained that conferences could assure that researchers got outside of the narrow world of academic life.[88] After citing Rousseau as an example of someone who was able to succeed outside of the academies, he admitted, 'others see in them [academies] a sort of mutual assurance society for scientific success; a system of reciprocal adoration'.[89] Additionally, Quetelet complained that the many academy proceedings were 'open to the public' and that debates were often influenced by the desire to sway the crowd. This kind of public science had been repudiated, and Quetelet hoped that conferences could allow scientists to gather in a private sphere, away from the need for ostentatious demonstration. Though Quetelet had believed in a strong *académie* in Belgium, it was only the beginning of his idea for associations. 'A conference', Quetelet claimed, 'will be advantageous in supplementing the insufficiencies of an academy'.[90] While 'their origin had been recent', Quetelet envisioned a future for large meetings of *savants*. The small group he found at Heidelberg was only the beginning.

Quetelet returned to Brussels in late October of 1829, likely satisfied with his mission. While the young Belgian writer of a decade before had found the German states to be a 'curious land', filled with a dark history and untrammelled nature, the Germany he witnessed on this trip must have seemed more familiar. It looked like his vision for Belgium. The observatories of Altona and Bremen, the academies and institutions of Berlin and the conference in Heidelberg were all examples of the kinds of science for which Quetelet had advocated in Brussels. German science, outside of Goethe and Humboldt, was becoming hierarchical, routine and standardized. The educational system was producing engineers and practical scientists, not eclectic dilettantes or idle speculators. Even the German bureaucracy itself had been developing the dedicated 'men of state' whom Quetelet admired. And with a supportive government and 'able generals' like Humboldt and Schumacher to lead the new bureaucracies, the future of the sciences in Germany seemed to be on secure footing. The Laplacian Empire of Paris may have represented the best example of scientific activity in the past, but Germany was the future.

Conclusion

Travelling in the German states convinced Quetelet that his vision of government-sponsored scientific activity was possible and that the German territories could be counted on to help in the effort to gather information in astronomy, meteorology, magnetism and population statistics. He had learned from Arago and Bouvard how one of the great eighteenth-century scientific institutions had operated, and he was a confident traveller in Europe after taking over sole editorship of an important continental journal. Even allowing for the generosity of Quetelet's own account of the trip and the generally congenial nature of the day, Quetelet's reception represented how much progress he had made in elevating Belgium (and himself) in the estimation of other European men of science. In interests and temperaments, one could find no more divergent group of scientific researchers than Bouvard, Arago, Villermé, Laplace, Poisson, Fourier, Schumacher, Olbers, Humboldt, Goethe, Gauss, Brown and Whewell, yet all of them responded favourably to Quetelet's ideas and enthusiasm, and each would become a regular correspondent.

Of course Quetelet did not get by on personal persuasion alone. By 1829 he toured Germany as the director of an observatory, an influential member of a thriving academy and the editor of an important European journal. As a man of science, Quetelet could point to these Belgian institutions as examples of his commitment to the international projects of science that would dominate the next half century. The fact that he had made little in the way of new discoveries was inconsequential. German men of science, as interested as Quetelet in expanding networks of data collection, recognized Quetelet as a valuable source of information. The success of the *Correspondance* in attracting large data sets rewarded their enthusiasm. After working for close to six years to reorient Belgian science towards the goal of international collaboration, Quetelet was in a position to share and receive a wide range of new data. He now had both a resource and an audience for what he believed to be a new science, one far more controversial and provocative than the astronomical and meteorological research he had been collecting so far.

5 *PHYSIQUE SOCIALE*, 1825–35

> What gratitude would not be due to the man of genius who set himself to draw up physical, metaphysical, moral and political tables, in where would be indicated with precision all the various degrees of probability, and consequently of belief, that should be assigned to each upon.
>
> C.-A. Helvétius[1]

Man of Genius?

Quetelet's work in the 1820s to create a European network of observers, statisticians and bureaucrats would not be realized fully until 1853, when he organized two international conferences in Brussels. Yet as Quetelet put in place his dream of national, continental and global conferences modelled on the 1829 Heidelberg conference, he never stopped investigating the applicability of probability and statistics to human behaviour – he never stopped pursuing *physique sociale*.[2] Developed during the decade he waited for an observatory, it remains his lasting legacy in the creation of the human sciences. It has assured him a place in most surveys of sociology and statistics, occasional mention in general histories of the nineteenth century and a moderate degree of recognition in disciplinary histories of criminology, magnetism and meteorology. This literature has largely seen Quetelet's work as a direct extension of Condorcet's plan for a science of man or as a 'social analogue' to Laplacian astronomy.[3] The story of *physique sociale* is not entirely the result of working out Enlightenment ideas, however, or the simple application of Laplacian rules to social statistics. As much as social physics was influenced by ideas of probability and progress, it was grounded in new forms of collaborative and professional data collection that marked a break with the Enlightenment understanding of scientific research.

Physique sociale should also not be interpreted as a set of ideas that were developed independently of Quetelet's other great passion for creating institutions for scientific observation. In fact, the opposite is true: Quetelet's social theory (which owed an explicit debt to the *philosophes*) was linked inextricably to his practice of science (which did not); Helvétius after all had asked for a *man* of genius to create table of probability, not an international network of trained observers. At almost every point along Quetelet's development of *physique sociale*, he relied upon data, theory and observations from his friends, colleagues and contributors. And at almost every point, the data flowed to Quetelet through his positions as observatory director, secrétaire perpétuel of the Académie royale or editor of the *Correspondance*. As suggested in previous chapters, it is quite possible that Quetelet pursued this research to further his institutional standing rather than the institutions serving a predetermined scientific end. It is also possible that for him such a separation between theory and practice would have been inconceivable.

Though Quetelet returned occasionally to *physique sociale* throughout his life, the focus of this chapter is the work done in the decade after his first paper on statistics in 1825. In 1835, the year the observatory finally received the expensive instruments from Paris and London, Quetelet moved on to data collection in magnetism, meteorology and periodic phenomena like shooting stars, and largely abandoned social statistics.[4] These projects were all significant in their own right, allowing even greater international cooperation among men of science and institutions, but they had only a limited impact on Quetelet's work in social physics. As Bouvard had predicted, the observatory would limit the directions in which Quetelet had worked, and Quetelet himself admitted that social physics was only a 'break' from the observatory project.[5]

Yet the relatively short period between 1825 and 1835 resulted in an incredible amount of work, only part of which will be treated here. After a brief summary of the state of social statistics as Quetelet found them, the chapter covers his first three papers on statistics, all delivered to the Académie royale, and all based on work Quetelet did prior to the 1830 Belgian Revolution while waiting for his observatory to be built. After 1830 Quetelet returned to his research in preparation for *Sur l'homme et le développement de ses facultés*, his most comprehensive effort to compile all of his research aims and the closest document to an articulation of his social theory. This time is also when he introduced in broad usage the two terms most associated with his ideas – *physique sociale* and *l'homme moyen*. It has been speculated that Quetelet's experience during the revolution led him to seek more order in his theory, but the post-revolutionary papers suggest other reasons why Quetelet's interests moved towards averages.[6] Quetelet's work can be situated in a broader intellectual and political context, but as this chapter argues, social physics was due at least as much to the particular and immediate force of the numbers on hand, collected through Quetelet's many networks of *savants*, as it was to larger historical events. The numbers did not quite speak for themselves, but in their presentation and reception within the language of Quetelet's unique network of scientific administrators, they received their most powerful spokesman to date.

The Father of Modern Counting

What does it mean to call Quetelet 'the father of modern statistics'?[7] He was certainly not the first person to aggregate numerical data for the purpose of drawing larger conclusions. In 1676, John Graunt had noted that his fellow Londoners made 'little use' of the Bills of Mortality drawn up by the city. Instead, he suggested they be:

> reduced into Tables so as to have a view of the whole together, in order to the more ready comparing of one Year, Season, Parish, or other Division of the City, with another, in respect of all the Burials, and Christenings, and of all the Diseases, and Casualties, happening to each of them respectively.[8]

From a mass of purely descriptive data, Graunt produced an elaborate set of tables along with conjectures about the numbers' meanings. Thirty years later, Johann Süssmilch found proof of divine law in the consistency in birth and death rates in his native Germany. During the eighteenth century, political economists in England became concerned with numbers, and in France Condorcet seemed to answer Helvétius's call for a 'man of genius' who would 'draw up physical, metaphysical, moral, and political tables'. In the world of probability theory, Quetelet was even less influential, with most histories of the discipline mentioning Quetelet as a propagandist at best.[9] Not only do most accounts of probability deny Quetelet fatherhood; he is left completely out of the family tree.

If the old paternity no longer applies, recent historiographical developments might best make the claim for Quetelet as the father of modern *counting*. Since the early 1980s, after near a century of neglect, the understanding of the development of probability and statistical techniques has received significant attention, and with it has come a much broader perspective on what has been called the 'probabilistic revolution'.[10] Although Isaac Todhunter's 1865 survey of statistics did an admirable job of explaining the early history of the mathematics of probability, the late twentieth-century literature expanded the understanding of what might be called a 'culture of probability', which developed nearly a century after the initial mathematical rules were created.[11] Such attention has refocused historians of science on a simple concept that had received comparably little consideration: how people in the sciences counted. The works have ranged from broad philosophical studies of the meaning of probability to technical studies describing the formal mechanisms through which Quetelet and others refined the act of counting.[12] This has resulted not only in more popular works on the nature of probability, but also a more profound understanding of what different means of counting meant for key ideas in the sciences, including such foundational issues as 'observation', 'description', 'experiment' and 'chance'. Though the literature is not in complete agreement on many issues, it has demonstrated that statistics was not a fully completed discipline which was *then* incorporated into the sciences – less still a technique used in the natural sciences that was then appropriated by the social sciences – but rather a complex set of techniques and ideas that 'emerged' from the interaction of interests in a number of fields and contexts.[13] Though figures like Quetelet clearly believed they were simply 'applying' ready-made techniques from astronomy, it was not so simple a process of diffusion. Governments, businesses and large scientific organizations did not have a ready-made discipline at their disposal, but instead had to create many of the tools they would employ over the course of the nineteenth century. Though a full review of the state of probability as Quetelet encountered it in 1825 would be too long for this chapter, a brief review of the state of statistics and probability up to this point is essential to understanding his project.[14]

Quetelet began his work at a moment of two related changes in the understanding of numbers. The first significant change was philosophical and linguistic; it was a change in what 'probability' itself meant. 'Classical probability' from the time of Pascal had been a relatively simple concept, meant to evaluate the chance that a certain future event might happen for a 'reasonable man'.[15] Developed as a means to wager on games of chance, it was concerned with what an *individual* in a given circumstance could expect. Pascal's foundational work on probability, for example, was written in response to a letter from his friend the Chevalier de Méré asking for assistance in splitting the proceeds of a game of 'points'.[16] Such a calculation was invaluable for Enlightenment authors who privileged individual choice and reason over collective and tradition-based decisions. Yet in the nineteenth century, probability began to move from a tool of the individual subject to a tool of objective oversight – from determining the correct choice one should make to a more determined view of inevitable outcomes. The difference in probability between *how likely something was to happen* and *how often it occurred* would have a profound effect on Quetelet's social theory, even as the Belgian helped create this transition.[17] Though he often tried to deflect criticism of deterministic language, the abandonment of classical probability left little room for a subjective theory of probability.

The second major change in thinking about numbers was political. In contrast with twenty-first-century states, governments by the early nineteenth century had few means of quantifying their populations' behaviour. Though bureaucracies had existed in many different forms and states had gone to great lengths to make their citizens 'legible', there had been only sporadic attempts at using numbers to extend state control.[18] Most bureaucracies instead followed the Prussian form of *statistik*, which was decidedly non-quantitative and focused on qualitative and descriptive means of understanding state service.[19] This was due in part to limited resources: as Quetelet's later work in Belgium proved, European states needed more people to conduct authoritative surveys. Practical limitations aside, governments also had to learn to think in terms of numbers and to view the traits of their populations as statistical aggregates rather than inherent qualities. As Stephen Stigler has noted, the *concepts* of probability had existed since the turn-of-the-century work of Laplace and Gauss, but they had not been adopted outside of a few physical sciences.[20] Quetelet, however, was the first to begin counting the data in state records with an explicit belief that what was found there would match the regularity of physical data. By combining the new developments in probability with a new interest in government in the actions of their populations, Quetelet became the first person to count human activity the way astronomers measured the stars.

When Quetelet first encountered the work of the Parisian mathematicians, he was in the perfect position to merge mathematical probability and state oversight, a combination that was decidedly non-classical. Astronomers, for example, did not use probability to determine what 'choice' a planet might make in its orbit around the sun, and Quetelet likewise saw *physique sociale* as a way to determine a projected course of society rather than as a way for an individual to navigate that course. Lorraine Daston has claimed that in Quetelet 'the transition [away from classical probability] was complete', and Quetelet was explicit that his science was of little use in helping particular individuals.[21] As seen in Chapter Six, the seeming effacement of individual choice attracted criticism, but it was a natural extension of the key contribution of *physique sociale*: the integration of statistics and *statistik*. In this vision, the individual was irrelevant because when the data were large enough the arbitrary factors of individual choice – what Quetelet called 'accidental causes' – were neutralized.

The propositions of *physique sociale* may have had parallels in the natural sciences, but they were also initially driven by the available data. Quetelet's early attempts at counting a population's behaviour relied on a few previously published sets of data in Belgium and Europe, and though he found the results interesting, he cautioned that much more work had to be done and that the existing data was hardly reliable. In the late 1820s, with the observatory still under construction and few means of acquiring data, he also had to rely on friends and subscribers to the *Correspondance* to provide most of the numbers. Though the early works did anticipate some of Quetelet's later conceptual advances, they also laid bare his need for greater administrative authority. The numbers would not come all by themselves, and here the father of modern counting made his most lasting contributions to the sciences of man.

The First Laws of Social Physics

In the five years prior to the revolution, Quetelet claimed two fundamental necessities for a science of man. The first was a better appreciation for probability theory. Though the 'calculus of probability' had existed for close to two centuries and 'served as the basis of study in all the sciences', Quetelet claimed that it had only recently taken hold: 'the *philosophe*, or at least the man who aspires to merit that title, must be able to estimate based on sure rules'.[22] In Paris, he watched Laplace and Poisson build observational astronomy on more solid foundations, and his natural next move was to apply the same techniques to studying man. The second step in uncovering the 'laws' was also based on the work of the observatory astronomers, but it required no new theory: researchers simply needed to be trained to acquire more information on what people did. Much as astronomers needed both new mathematics and new observatories to make new discoveries, the social physicist needed similarly large networks of observers for the field to progress.

Quetelet notably began his statistical work in 1825 by addressing the second of these two goals: a summary of administrative data. On 25 April, he presented to the Académie royale his 'Mémoire sur les lois des naissances et de la mortalité à Bruxelles', a paper included the next year in the academy's *Nouveaux mémoires*. Despite the intent to explain social 'laws', Quetelet was careful to claim provisional status for all of his conclusions. There simply had not been much data collected, a point underscored by the fact that the best available records of mortality data in all of the Netherlands had been collected on annuitants in a small town in 1742.[23] Quetelet rarely relied on life insurance tables in his later work, and their inclusion in this paper indicated that the first step in creating a statistical account of society was to get more data. To supplement Kerseboom's data, he collected eighteen years of information from the town registry of Brussels. Aided by his former student Charles Morren – 'who had the patience to do the majority of the work' – Quetelet offered to the *académie* what he believed was the first large-scale collection of birth and mortality records in his country.[24] The results appeared to be a revelation.

Within the mundane entries of the city registry, Quetelet found his first 'laws' of society. They were modest conclusions, but statistically constant: births tended to be lowest in July (an average of 9,012) and highest in February (11,560); between these two months, the average number of births either increased or decreased continuously and, most importantly, the progression of these rates could be expressed mathematically so that the data fit a normal sine curve.[25] If the monthly data were to be plotted graphically (which Quetelet did not do in this paper), it would have looked much like what we think of as a sine wave, with the periodic peaks and valleys alternating around an average month. Joseph Fourier had begun to apply the sine curve to all sorts of phenomena, and Quetelet counted it a major success to express human behaviour in the same geometrical and analytic form that expressed heat transfer and asteroidal orbits.

Quetelet became even more intrigued when he compared the number of monthly births to monthly deaths. Throughout a period of seventeen years,[26] Quetelet observed a near perfect relationship between births and mortality in Brussels, with deaths similarly bottoming out in July and reaching their peak in February.[27] Not only did this mean that mortality could also be expressed with a *sinusoïde*, there might also be some possible relationship in society between the two primary movers of population change: births and deaths. Indeed, Quetelet already had a theory of society that might account for such a relationship. In Thomas Malthus's famous theory of scarce resources, he had attempted to show the 'vice and misery' that had resulted when population growth 'increase[d] beyond the nourishment prepared for it'.[28] The close *rapport* Quetelet found between births and deaths, he wrote, 'accords very well with the remark from Malthus that the number of births increases when there is a gap [*vide*] in the population'.[29] Though it might seem odd that mothers could plan their pregnancies nine months in advance in order to fill some future 'gap' in the population, the relationship makes more sense when the high level of infant mortality in the early nineteenth century is considered. In conditions in which many children failed to make it through their first days of life, births and deaths would almost certainly occur in similar frequencies. Quetelet also believed that it was no surprise that births and deaths should have such a strong relationship; they were both products of the same law which produced the seasonal regularities and therefore provided evidence for a Malthusian theory of population.[30]

As interesting as the regularities in monthly data were, Quetelet understood that they could be explained by seasonal fluctuations: winter months obviously made it more difficult for children and weaker adults to survive. While discovering a link between climate and population change was exciting in and of itself, it was not enough to indicate an autonomous law. Quetelet understood that the law observed was the result of the seasons and therefore an extension of the physical laws of planetary movement and axial tilt. Regularities in births and deaths then were nothing more than a tertiary expression of Newtonian laws. The birth and death rates, shocking as they were in their consistency, were not sufficient to conclude that independent social laws existed.

Quetelet found a less dependent law, however, when he decided to group mortality data according to age rather than months. Combining all the data, he focused on the number of deaths as well as the total living population at each age. The ensuing three-page chart was stunning. In a table labelled 'Loi de la mortalité' (see Figure 5.1), Quetelet listed the total number of deaths for those in their first year of life (14,261) all the way up to age 102 (one). The fact that more people died in their first year than at age 102 was hardly surprising, but as the table showed, the progression from zero to 102 was *continuous*, meaning that the number of deaths decreased at each age: fewer people died at age nine than age eight, age forty-eight than forty-seven, age sixty-two than sixty-one and even at age one hundred than at ninety-nine! Were the number of deaths plotted on a graph with the number of deaths as the y-axis, and year of life the x-axis, it would have been a descending line from age zero to 102, without a single increase. Even in the first year of life, the numbers mostly fell month to month, from 1,044 in the first month of life to 142 and 140 in the eleventh and twelve months.[31] The same numbers held true for the total population at each year of life, which Quetelet listed in a table entitled 'Loi de la population'. While a general decrease in mortality and population was to be expected, such unvarying regularity suggested to Quetelet that phenomena as individual, and at times seemingly random, as death and total population offered quantifiable frequencies as predictable as any found in the observational sciences. In his words, 'we are authorized to believe that the laws ... found in nature in the development of plants and animals ... must be extended to the human species'.[32] With just one assistant and a few years of data from his adopted home of Brussels, Quetelet believed that he had already confirmed that the laws of social physics were real.

LOI DE LA MORTALITE A BRUXELLES.

AGES.	HOMMES.	FEMMES.	AGES.	HOMMES.	FEMMES.	AGES.	HOMMES.	FEMMES.
0	7418	6843	35	2921	2977	69	904	1167
1	5674	5536	36	2873	2928	70	835	1096
2	5023	4942	37	2824	2879	71	767	1023
3	4654	4614	38	2774	2830	72	699	948
4	4431	4409	39	2723	2780	73	631	872
5	4304	4225	40	2671	2730	74	564	797
6	4194	4209	41	2618	2680	75	498	723
7	4138	4137	42	2564	2629	76	433	650
8	4089	4100	43	2509	2578	77	369	581
9	4051	4069	44	2453	2527	78	319	517
10	4026	4039	45	2396	2476	79	269	457
11	4005	4016	46	2338	2425	80	234	402
12	3986	3992	47	2280	2374	81	202	352
13	3968	3967	48	2222	2323	82	172	307
14	3951	3941	49	2164	2272	83	144	268
15	3935	3914	50	2105	2221	84	119	232
16	3915	3887	51	2046	2170	85	97	198
17	3893	3859	52	1987	2119	86	76	165
18	3863	3830	53	1928	2068	87	54	133
19	3826	3799	54	1869	2017	88	42	101
20	3781	3766	55	1809	1966	89	30	73
21	3714	3721	56	1749	1915	90	21	55
22	3647	3673	57	1689	1864	91	14	41
23	3581	3620	58	1629	1813	92	9	30
24	3518	3563	59	1568	1762	93	6	22
25	3455	3505	60	1506	1711	94	4	17
26	3394	3448	61	1444	1659	95	3	14
27	3333	3392	62	1381	1606	96	2	11
28	3273	3337	63	1316	1550	97	1	8
29	3218	3283	64	1249	1492	98		5
30	3166	3230	65	1181	1431	99		2
31	3116	3178	66	1112	1368	100		1
32	3067	3127	67	1042	1303	101		1
33	3018	3077	68	973	1236	102		1
34	2969	3027						

Figure 5.1: Loi de la mortalité. As seen in this table, the number of people who died at each age decreased continually throughout life. Quetelet was struck by how consistent the pattern was, with not even a single age having an increase over the previous year of age. Correspondance mathématique et physique, 1 (1825), p. 218. Courtesy of Boston Public Library Rare Books Room.

The Problem of Numbers

From his experience with the Parisian mathematicians, Quetelet knew the conclusions he reached in his 1826 'Mémoire' needed to be confirmed by other observers before they could be considered reliable. Furthermore, he would have to look at human behaviour besides mortality and procreation. Fortunately, the 'Mémoire' came out just a year after he became co-editor of the *Correspondance mathématique et physique*, which allowed him access to research outside of the Low Countries. In a note in the first *Statistique* section (which also included a summary of his Brussels data) Quetelet made an appeal to his subscribers to provide him with more information. He believed that it was possible to 'disentangle' some 'general laws' from an accumulation of information and that the acquisition of more information would 'further extend the circle of [our] knowledge'. For now, Quetelet explained, he could only present what little work he had done so far and simply note that he was 'struck by the regularity'.[33] The rest of the effort would have to be undertaken by his correspondents.

The readers of the *Correspondance*, mostly administrators in the United Kingdom of the Netherlands and in France, answered Quetelet's call with a wealth of information from 1825 until 1827.[34] The resulting work, published in 1827, but including material already published during the previous three years, was a tour de force of quantitative statistics. Though Quetelet was listed as the sole author, it announced a new level of collaborative research. In *Recherches sur la population, les naissances, les décès, les prisons, les dépôts de mendicité, etc.* (1827), Quetelet applied everything he had learned as a student of probability to the multitude of observations he had received from continental researchers and administrators. Comparing the numbers, Quetelet not only confirmed his 'loi de la mortalité' in other countries – the number of deaths also fell continuously in Amsterdam and Paris – but also presented a host of esoteric data, including such bizarre trivia like the fact that in an average year in France, there were 138 people living for every marriage performed.[35] At close to ninety pages, Quetelet covered all of the elements of society found in the title as well several digressions on the possibility of applying probability more broadly into politics, morality and philosophy. The result of the *Recherches* was a hodgepodge of evidence, enthusiasm and tentative conclusions.

Though the richness of Quetelet's 1827 work allowed multiple interpretations, from the perspective of Quetelet's later social physics, there were three key elements to the work. The most important idea was that Quetelet expanded upon the meteorological metaphors invoked in the 1825 paper. After explaining that evidence garnered from Lobatto's *Jaarboejke* indicated that all the 'qualities in the different provinces of the Kingdom' could be fitted to a geometrical curve, he claimed that 'this method is close to that adopted by *physiciens* to express variations in temperature and barometric pressure'.[36] Furthermore, before comparing his own records on Brussels with what Louis-René Villermé had found in Paris, he wrote that 'the following numbers are a little like the variations found in a thermometer'. Such a relationship – that both climate and population could be expressed in the same graphic and analytic form – led Quetelet to believe that there was a deeper connection between the two. The relationship was, 'in a sense, that of opposites: that is to say that at the times when the number of degrees on the thermometer scale is the highest, the numbers of births and deaths is the lowest'.[37] Conversely, in the winter months when temperatures dropped, mortality rose.

The relationship between a given climate and the 'qualities' of a population led Quetelet to explore causality, the second key feature of the *Recherches* and one that would remain a constant in his career. In Quetelet's first work, he had limited discussion of causation and was content to present a few tables and describe the changes in broad strokes (Quetelet's 1825 'loi de la mortalité' did not *cause* the mortality rate to continuously fall; it *was* this continuous fall). After all, Quetelet believed he was basing his laws on the physical laws described by Laplace, and it would not have made sense to say that something like inertia *caused* bodies to remain in motion or that seasonal shift *caused* temperature changes. Astronomers and physicists did not ask what *caused* inertia or gravity.[38] Yet in *Recherches*, Quetelet had access to far more than mortality rates, ages and seasons and was therefore able to posit a more direct relationship between different phenomena. The first connection he found was among physical location, climate and mortality in Lobatto's numbers from western Flanders: 'it strikes the eye ... that the provinces most populated and closest to the sea are the most exposed to mortality'.[39] Even more suggestive was what he learned from his correspondent Villermé's comparison of birth rates in Paris, Palermo and Florence. While Villermé's Paris had a similar climate to Brussels, and therefore showed a similar peak in birth rates in July, Palermo and Florence saw their birth rates spike in June. As Villermé reminded Quetelet in a note to the *Correspondance*, the fact that the numbers showed similar trends and different peaks was evidence that it was the temperature that mattered most.[40] Agreeing with his colleague, Quetelet suggestively asked: 'from the previous [data], must one conclude that the temperature is effectively a direct cause of greater or lesser fecundity?'[41] Such a relationship between climate and human activity was hardly new and had been a feature of French social theory since Montesquieu, but for Quetelet it acquired new force because both temperature change and birth rates could be quantified and expressed as easily as the most immutable physical laws. An emphasis on strong causal language and a naturalistic metaphor would eventually lead to charges of determinism against Quetelet; after all, the critics claimed, if the planets could not choose to alter the path of their orbit, nor the seasons their progression, how could people *choose* something as quantitatively consistent as the decision to have children?

Quetelet later adopted more deterministic rhetoric, but at this point he offered a nuanced picture of how social 'laws' influenced the free will of human actors. Perhaps the best explanation of Quetelet's ideas at the time came from a discussion on why it was that procreative sex dropped in the coldest months. The relationship on one level was obvious: the hardships of winter subdued sexual desire. In Quetelet's words, it was 'the rigors of winter' where 'people were forced to struggle against the intemperate air'. Yet such privations did not directly 'cause' the suppression of sexual activity; rather the hardships of winter operated through the function of lowering a person's 'morale' to the point where they were able to make an unexpressed decision not to reproduce: 'it is the pain which influences the morale of man, and by extension his physical faculties ... it excites him less to reproduce'. While this may seem to be a direct relationship, it did not deny the presence of an individual will in the matter – only that the 'rigors of winter' were so strong that they influenced conscious thought.[42] As seen in the following passage, Quetelet suggested that while there might be a decision ('calculation') by an individual subject, this was not the same thing as free will.

> Before having even felt the consequences of his actions, by a natural calculation of which his will perhaps does not even take any direct part, a man finds himself stopped by the fear of adding to his misfortunes by seeing his family grow.[43]

What, then, was the 'natural calculation' if it did not involve a will in a 'direct' role? It would seem to be the act of an individual realizing a social law, which for Quetelet would make cognition something similar to the biological process which caused plants to 'know' when to grow in the spring. As deterministic as it may seem, however, such a suppression of will could be overcome. It was not that the meteorological forces left no room for free will, but that climate merely restricted an individual's capacity to act on his will. Quetelet chose his words carefully, using most often the French *entraîner* to describe the relationship between the weather and human action, with the verb meaning either 'to cause' or 'to carry something along' in English. Furthermore, in describing the effect of climate on the death of young children, Quetelet fully acknowledged that climate was secondary, quoting the French statistician Benoiston de Châteauneuf that 'in order to preserve children's lives, care is everything, and climate nothing or very little'.[44] A parent's love for a child, presumably an act of will, could steady the influence of the seasons.

The third novel element of the 1827 *Recherches* was that Quetelet finally was able to examine statistics outside of fecundity and mortality such as poverty, education and crime. Here was the chance to find a true social physics, one that could explain human *actions* rather than simple biological features of existence like birth and death. While people rarely chose when they died or in which season they procreated, if laws could be found in seemingly freely willed actions like employment and criminal behaviour Quetelet would be closer to finding expressions of laws embedded in society rather than the weather. Unfortunately, Quetelet's initial investigations into poverty, education and crime seemed to lead to data that were irrelevant or incomplete. In analysing the populations of the *dépôts des mendicité*, Quetelet believed the numbers 'inspired confidence' and would 'certainly be of interest to the friend of humanity who interests himself with the unfortunate of society'.[45] Yet there was very little in the numbers to indicate underlying causes of poverty. Other than to note that men and women appeared in the *dépôts* with equal regularity and that there existed a large variation among regions, Quetelet's only analysis was to point out the 'modestly frightening relationship' between poverty and mortality. The inhabitants of the *dépôts* were more than four times as likely to die as those on the outside. Such a relationship accorded well with the data he received from his regular correspondent Villermé, who had posited a relationship between wealth and mortality, but Quetelet cautioned that the numbers needed to be analysed region by region, because the *dépôts* in distant cities like Rekem (on the German border) and Bruges (by the North Sea) showed almost no relationship. However, even here, Quetelet hesitated to say anything definitive because 'the number of observations for these two places [was] too small to have any confidence'.[46]

The lack of confidence in the numbers was especially acute when it came to education. To his dismay, Quetelet found that there were more than double the number of schools relative to population in the 'north' (the modern Netherlands) than in the 'south' (modern Belgium), a striking difference that called for some explanation.[47] Even worse, how to explain that his beloved hometown of Ghent, which 'some [had] rather pompously titled the *Modern Athens*', revealed such low rates of attendance? After dismissing the possibilities that his native Catholic provinces cared less about education and that these institutions were so superior that they could educate twice as many students, Quetelet came up with a third possibility: 'there is an error in the numbers'.[48] He was quick to point out that Belgium had many alternative schools found in churches and that these would not show up in the official numbers (though he did not note that it was his patron King William who was responsible for closing so many of these schools). In short, Quetelet declared that 'one must therefore conclude with circumspection the results which have been obtained by the government'.[49]

Quetelet faced further difficulty in government numbers when it came to criminal behaviour, the subject that would preoccupy much of his own work as well as the future field of sociology. After first decrying the 'great number of prisons' in the country, Quetelet asked if the prisons were doing any good, either for society or for the prisoners themselves. Certainly the data revealed that there were far too many citizens in jail, but there was little else that could be gleaned from such raw data. Recidivism rates were unknown, and Quetelet could not even determine the mortality rate in prisons because the numbers were unavailable.⁵⁰ All he could see were the crimes for which the prisoners were incarcerated, knowledge that 'merited the attention' of any man 'who wished to make disappear the scourge which damages so much of society'.⁵¹ Yet little could be done to remove 'the scourge' of crime without better numbers.

Quetelet's wish to make crime and poverty disappear dispels the idea that social laws pre-determined all human action. His conviction was that social physics could not only uncover laws of human behaviour, but could also help governments modify the expressions of these laws. Hence once of his most famous comments, eight years later in *Sur l'homme*, that if governments embraced social physics it would be possible to decrease significantly the 'budget [of] ... the scaffold, the prisons, and the chains'.⁵² In a brief conclusion (a single paragraph after sixty-eight pages of data), Quetelet here only speculated that the people of the Low Countries were growing at a sustainable rate and that 'they could procure work for themselves'. Always mindful of Malthusian concerns, Quetelet called the growth 'very sensible', and the data confirmed that 'the means to multiply the population of the state ... is to facilitate the means of production in agriculture and industry'. Statistics, he argued, was the way for governments to maintain order and stability. Yet he warned that the population could 'not be determined with all desirable precision' and that governments owed it to themselves and their citizens to keep track of all activity relating to population. 'To encourage reproduction without assuring people of the ability to augment their own subsistence', Quetelet warned, 'is to impoverish the state and favour death'.⁵³

Quetelet's focus on the importance of uncovering statistical laws for the purpose of government action revealed that by 1827 he had found the two foundations – one intellectual, the other practical – on which to build his new science of social physics. The first was that social physics needed to establish causal relationships among human actions. Relationships with external regularities like the seasons may be interesting, but researchers needed to go further and explore how constants, like temperature, affected mortality *through the means of its effects on behaviour*. The cold weather did not simply kill the child, Quetelet argued; it acted on the will of the child's parents to be able work and provide food and shelter. This meant, therefore, that what we would now call correlation was insufficient; there needed to be some descriptive mechanism that explained how one force acted on another, not just an overlap in numbers. The second foundation, the practical one, involved convincing governments to support the creation of large-scale statistical accounts of their population, lest they fail to notice imbalances in their population and subsequently 'favour death'. Quetelet's first sustained treatment of *physique sociale* was optimistic but guarded. The great minds of physics and astronomy had delivered a mathematical guide to creating stable societies, but it was up to governments to do the counting.

A Growing Confidence

While the 1827 *Recherches* was hesitant about the possibility of governments implementing the practical foundation, by 1829 Quetelet had become an unapologetic supporter of the intellectual goal of uncovering laws of causality within society, as enthusiastic as Condorcet about the power of quantitative reasoning to secure a better future. In the preface to 'Recherches statistique sur le Royaume des Pays-Bas' (1829), the third and final of his pre-revolutionary writings, Quetelet offered a powerful summary of how the Enlightenment dreams of a social science could be put into action through statistics and probability.

Like many of the *philosophes*, Quetelet believed that until the eighteenth century the history of human actions had been treated in broad and sensationalist terms that did little to explain what motivated historical actors. Voltaire and d'Alembert had argued for a more 'philosophical history', replacing the deeds of great military and political figures with the ideas of great thinkers, and Quetelet echoed the complaint of a tradition where history writing was all lurid spectacle.[54] In this vision, Quetelet wrote, human history was 'reduced to a portrait of the deplorable effects of [humanity's] rage'. Seemingly without cause, men were carried on a 'current of bloodlust' or 'blindly serving the instrument of vile passions'. But due in large part to the new forms of history of the seventeenth and eighteenth century, there was now a more 'consoling form of study', where mankind was 'considered more closely' and 'where one can search to penetrate the secret of their riches and survey [*sonder*] the sources of their prosperity'.[55] For Quetelet, *physique sociale* was a similar way to 'survey' the present, looking beneath the surface of events to uncover deeper patterns and regularities.

The key tool to penetrating the depths of social phenomena was again probability. It had proved its worth in examining the hidden movements of physical bodies, and there was good reason to suspect similar hidden explanations in society. While *statistik* had given qualitative accounts of the state, in the future statistics had 'to follow on the same road as the sciences of observation' through implementing quantitative measurement.[56] Probability theory was valuable because it had shown in physics and astronomy that seemingly inexplicable events had underlying causes, and in the case of Laplace, had helped shape the laws of physical bodies. Just as Laplace had removed the necessity of God in Newtonian laws, similarly Quetelet wished to dispense with the secular equivalent of divine intervention: random chance. In his words, 'we must rob ourselves of the notion' that because one does not see direct causes, there must in fact be none. Moreover, *le hasard* was a 'word lacking in sense' in 'which the vulgar hide their ignorance'. This ignorance was inspired by 'the vague and foundationless hypotheses and systems' which had had subverted true knowledge. 'In place of words', Quetelet wrote, echoing a century of Enlightenment thought, 'one wishes for facts'.[57]

As Quetelet's earlier writings show, the fledgling field of social physics had barely begun to take advantage of probability theory because, in contrast to the observational sciences, his science had no network of observers to compile the data. By doing away with what he called 'vague hypotheses', Quetelet also wished to banish the speculators themselves. He believed that the causes of social phenomena were so complex that no single person could understand them all.[58] Rather, in order 'to show the effect of causes, and to conclude that which is and that which will be', social physics required a large system of observers to compile accurate data. Like many later critics, Quetelet recognized that statistics could be made to say anything with poor data. But this was a criticism of the field as it was, not as it could be. What would physics, astronomy and chemistry have been like, Quetelet asked, if their fields had been dismissed because 'observation was difficult' or because some of those who practised it 'were ignorant men or those of bad faith?'[59] Like the more mature sciences he wished to emulate, social statistics needed time to find men of better faith. As his work in Belgium and Europe to establish networks shows, Quetelet had a good idea where to find these new men.

The preface to 'Recherches statistique' made explicit Quetelet's primary criterion for a *physique sociale*, one he had only hinted at in the prior two papers. In addition to quantitative analysis done by learned probabilists like himself, the science needed reliable numbers produced by trained observers. Using probability continued in the tradition of late Enlightenment thinkers, but the data collection constituted a new programme. It required not only mathematicians to extend the work that had been done in probability but also required new investments in interchangeable observers who could count social behaviour. Quetelet's three prerevolutionary writings showed *physique sociale* to be on one hand a faithful model of an Enlightenment vision for a science of man, based in the mathematical laws of observational science and useful in improving humanity's material wellbeing. On the other hand, the plan for social physics required the development of a new form of observer, one very different from the eighteenth-century *savant*.

In spite of the deterministic hints and the problem with faulty numbers, the early data were for Quetelet a powerful sign that social physics was possible. With only incomplete data and a limited number of correspondents, Quetelet had in the span of four years done more to connect probability theory and *statistik* than any previous thinker. He had discovered a 'loi de population' that, while echoing Malthus, had grounded the theory in better numbers and more elegant mathematical formulations. After all, Malthus had influenced public policy by relying on a crude extrapolation from limited data; what an improvement Quetelet believed it to be to find an expression of mortality and birth rates that resembled a sine wave. At this point, social physics seemed to be a real possibility, revealing a social world every bit as neat and predictable as the motion of heavenly bodies.

Why Average?

Of course, as Quetelet was always ready to acknowledge, individual human beings had an annoying tendency to disrupt social laws. One of the predictable features of society was that it was unpredictable. In Quetelet's case, the unpredictable occurred in August 1830, when *les Bruxellois* stormed into the streets to protest the reign of the king Quetelet had long courted. Here the unpredictable individual may have been Adolphe Nourrit, the fiery tenor whose performance as Masaniello at the Théâtre Royale de Monnaie drove the crowds out into the Brussels streets on a hot August night. Or perhaps it was Daniel Auber's riveting score of *La Muette de Portici*, which had so excited Parisians just a month prior. Or perhaps it was William, the king of the brief polity of the United Kingdom of the Netherlands, who had been too aggressive in pushing a Dutch model on the Catholic provinces. Or maybe it was Quetelet himself, who had done so much to help foster ties between Belgium and the French. Or maybe the unpredictable could be found in the hundred or thousand other actors that led to the creation of a new country of Belgium in 1830. In any case, the myriad factors behind the Belgian Revolution posed a problem for Quetelet: how could the events of the summer of 1830 be *explained* by a social physics, let alone predicted?

The Belgian Revolution was far from an abstract political concern in his life. Quetelet was in Italy that summer and had left his wife Cécile and two children in Brussels. Contact outside of the city was difficult, and Cécile was forced to ask Quetelet's friend Bouvard for help. The director of the Paris Observatory had offered his assistance, but by the time he could get word to Cécile, she had made other plans, taking the family first to Ghent and then Lille before returning to Brussels two months later. In addition to not knowing the condition of his wife, Quetelet had many friends and family involved in the affair and the foundation of his observatory had been used as a military barracks (it was, after all, the highest point in Brussels).[60] Therefore it might be reasonable to suspect that the ideas he developed immediately afterwards – including 'perturbing causes' and *l'homme moyen* – were due to a concern about social unrest. Such arguments have been made, and there are strong connections between the kind of equilibrium Quetelet looked for in his statistics and the disruptions of 1830.[61]

While the revolution caused Quetelet significant trouble, a review of the pre- and post-1830 writings on social statistics makes it hard to see how *physique sociale* would have differed appreciably given another series of political events. And as important a moment as it was, political revolution was hardly unprecedented in Belgian life; Quetelet's friends and family had already experienced turmoil during the French occupation of Belgium, and Quetelet was nineteen when the great disruptions caused by Napoleon ended in Waterloo, just eighty-six kilometres south of Ghent. Nearly all his good friends had fought either for or against the French general in the most 'perturbing' events in Europe in centuries. Quetelet's immersion in the Parisian sciences and his interest in d'Alembert, Condorcet and Diderot gave him a full appreciation of the revolutionary years in France. Furthermore, prior to 1830 Quetelet had already acknowledged that unprecedented events could cause disruptions to his laws of society, specifically in his population data when he excluded the *annus horribilis* of 1815 from the mortality tables. Finally, the observational sciences that Quetelet believed served as the inspiration for social physics had long recognized the idea of 'perturbing forces'. As he wrote in 1831: 'I name the perturbing forces of man by analogy with the perturbing forms that the *savants* have considered in *le système du monde*'.[62] If Laplace had incorporated 'perturbing forces' into celestial mechanics, Quetelet did not need the Belgian Revolution to incorporate them into *physique sociale*.

The social physics that Quetelet developed after the Revolution looked very similar to what he had been working on before. His 1835 book *Sur l'homme* incorporated entire sections of his pre-1830 papers and while the terms 'natural and perturbing causes' were new, the concepts were not. Quetelet had already explained that social laws were unlike physical laws because of the ability for human will to intercede, such intercessions being the very point of doing social physics. The Belgian Revolution, disruptive as it was, was certainly not a historically unprecedented event. Rather it was another anomaly in the social world, another extreme bit of deviance that occurred at the outer spectrum of possibility. Quetelet's work in the years after 1830, including three important studies that would be incorporated into *Sur l'homme*, could be imagined without any political revolution. The events of 1830 were not a refutation of the possibility of determining social laws, but an example of the fluctuations inherent in a probabilistic account of social action.

If it was not the revolution that caused Quetelet to develop *l'homme moyen*, his most famous and enduring idea, how could its appearance in 1831 be explained? The answer may be that Quetelet's thought was developing exactly as he had predicted: new data was leading to new conclusions. In fact, after the revolution Quetelet produced his first analysis of physical features, data which tended to lead towards the consideration of averages. *Recherches sur la loi de la croissance de l'homme* (1831), dealing primarily with height and weight data from Brussels and Paris, allowed Quetelet to see numbers that grouped themselves around an average in ways not found in population statistics, birth records or mortality tables. In this paper, Quetelet first deployed the use of the terms *minimum, maximum* and *moyenne* to describe the variations in physical characteristics, reflecting the patterns in the numbers he saw.[63] So uniform were the numbers from Brussels that Quetelet could eventually create a mathematical formula that predicted numbers outside of the data set. For example, while the observed *moyenne* for the height of people over age fifteen was 1.549 metres, Quetelet 'predicted' an average of 1.546 simply using the formula he had developed for those under the age of fifteen (for his later graphical demonstration of the average, see Figure 5.2). The formula could also predict the size of younger children and, to Quetelet's surprise, predict even the size of foetuses, finding that the observed sizes matched closely the ones he had estimated.[64] Unlike any other numbers Quetelet had seen, physical characteristics seemed to be amenable to the rules of quantitative interrogation deployed in the physical sciences.

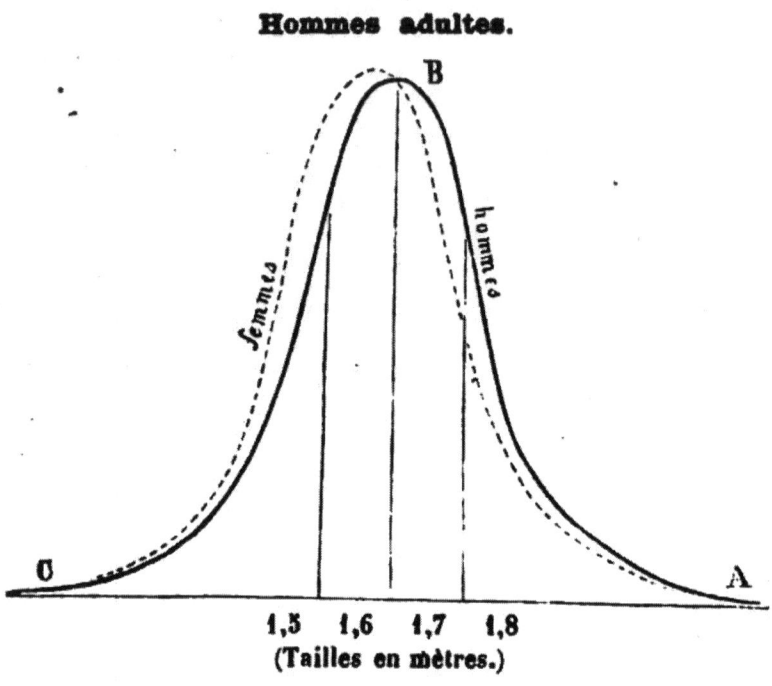

Figure 5.2: Qualités physiques. When Quetelet first started investigating tables of height in 1831, he noted that heights were grouped in such a pattern that he could 'predict' average heights for certain ages. Though Quetelet did not yet present the information as the 'bell' curve seen here, the distribution of the heights convinced him that human physical characteristics might follow the same laws as physics and astronomy. A. Quetelet, Physique sociale ou essai sur le développement des facultés de l'homme, 2 vols (Brussels: Muquardt, 1869), vol. 2, p. 22. University of Wisconsin Memorial Library.

In *Sur l'homme* in 1835, Quetelet built on these initial conclusions from the 1831 paper and incorporated a far larger amount of data on physical characteristics, all of which were similar to height in that the observations tended to group around a central point. The new data not only confirmed that certain observations occurred more frequently, but also that occasional anomalies were distributed evenly as the observations moved further away from the average. As he claimed about weight:

> If one were to place to the right of the man of average weight all the other men who weigh more than him, and to the left all of those who weigh less, in lining them up proportionate by weight ... one would be stunned to see the regularity which is established.[65]

Rendered in visual form, this description would look exactly like a bell curve. Quetelet did not claim the full connection between the distribution of sizes and the Gaussian error curve of measurement until his 1844 essay on probabilities, but the physical characteristics had started him in this direction.

This was another confirmation that *physique sociale* was possible at the level of physical traits. Height and weight could be expressed mathematically in the same way as natural phenomena, and therefore physical characteristics were subject to the same sort of laws as any other kind of data from the natural world. Like temperature readouts or determinations of star positions, height and weight followed the Gaussian distribution of error, where observations appeared less frequently as one moved away from the most common. Furthermore, tables of height and weight proved that accurate predictions could be made about unknown data using the minimum, maximum and average values as constants in a simple formula. Whether it was the error involved in observing celestial data or the natural 'error' of fluctuations in temperature, both 'distributions' appeared similar to those for height and weight.[66]

The importance of this consistent regularity between height and weight has not abated, and the regularity remains impressive. The durability of the regularity explains why this section also included a brief formula that unexpectedly became Quetelet's most cited 'discovery'. First presented to the Académie royale in 1832, his paper entitled 'Recherches sur le poids de l'homme aux différens âges' gave the world the Quetelet Index, or what would later be known as the body mass index.[67] Here Quetelet reasoned that if both height and weight were subject to similar laws in their development, and both showed relationships to age, sex and geographic region, it was natural to conclude that there was some relationship between the two. The relationship could not be too simple, because height and weight did not increase in the same *rapport*. This would be true only 'if man grew equally in all his dimensions' at the same time.[68] Instead Quetelet saw that from the first year of growth until puberty a relationship could be drawn between the square of the weight and height taken to the fifth power. This was helpful, but quantitatively cumbersome. Fortunately (or not) for countless doctors, public health officials and conscientious weight watchers, the relationship between height and weight in *adults* was far more elegant:

> [In comparing] individuals entirely developed and regularly constructed ... we find that the weight of individuals of different height are nearly like the square of their height.[69]

While Quetelet quickly moved on without further comment to examine how weight differed according to geography, this brief formulation would become his most cited legacy in the second half of the twentieth century and the twenty-first century. When later examined alongside health data, the Quetelet Index became the most consistently used and trusted measure of obesity and a staple of government health records. For the practical-minded Quetelet, he would have been happy to see the BMI in public health initiatives today.[70] At the time, however, the mathematical relationship between height and weight was simply another example of the power of averages.

Overcoming Absurdity: The Average Man of Intelligence and Courage

Had Quetelet stopped here, he would have left future generations with only the BMI measure of obesity and an interesting account of the interrelationship among physical characteristics. Yet what could be considered 'social' about height and weight? Like his early investigations of mortality and population growth, physical characteristics were largely outside of the individual's control and, therefore, closer to natural phenomena than they were to human and social behaviour. The reason Quetelet had chosen this data, however, was because it was what was available at the time; it was what early nineteenth-century European institutions had decided to count and measure. The next and most challenging step in advancing *physique sociale* was to expand the range of phenomena that could be *counted* to include moral and intellectual acts.

The first obstacle for a true social physics that measured moral and intellectual behaviour was to determine a unit of measurement. Age, height and weight all had obvious and generally accepted units, applicable to individuals and able to be aggregated, but how to count moral and intellectual ability? Quetelet began to answer this question by insisting that some moral behaviour could be counted, at least at the level of the state. Criminal activity, for example, was widely available, and while it could not be presented as an individual unit, the entire crime rate could be adduced by state governments and then projected back onto the general population. One might not be able to measure an *actual* individual's propensity for crime in a way that their weight can be measured, but the numbers existed to estimate a single individual's criminal propensity based on aggregated data. But what about moral or intellectual traits that have no quantitative correlate to criminal records? Could you really say that someone, even a hypothetical someone like Quetelet's average man, had a defined quantity of bravery, compassion or intelligence? Quetelet recognized the problem of quantifying these characteristics as early as 1831, and his objections to quantifiable moral and intellectual behaviour were as strong as any later critic:

> How can one ever sustain without absurdity that the courage of one man is to that of another like 5 is to 6?

And on intellectual qualities:

> Does one not laugh at the pretentiousness of a geometry which seriously believes that it can calculate the genius of Homer is to that of Virgil as 3 is to 2?[71]

The absurdity of quantifying courage and intelligence highlighted a second problem in measuring moral and intellectual characteristics: such traits could only be measured through *acts*. Unlike height and weight, whose units of metres and kilograms essentially *were* the physical characteristics, moral and intellectual behaviour could only be measured indirectly. As Quetelet would later write in the introduction of book four of *Physique sociale* in a section entitled 'Development of Intellectual Faculties': 'One can only appreciate faculties by their effects; that is to say by the actions or the works that they produce'.[72] Unlike a person's height, which is fixed and can be known without someone 'acting' on their height, a moral trait like courage could only be determined if a person *does something*, and then only counted if the means existed to observe and record that action. Cause and effect had entered into physical data as well, but causes (climate, geography, wealth) could easily be separated from effects (population, height, weight). In moral and intellectual matters however, cause and effect were not so easily separated – was there some way to actually distinguish between the 'cause' of inherent courage and the 'effect' of a courageous act? Deciding on a unit of measurement might be the easy part; finding something to actually measure was the real difficulty.

In 1831 Quetelet offered only a partial solution to the first problem of units and largely ignored the second, a mistake that caused a considerable backlash over the next decade as he defended himself against charges of determinism. To overcome these objections, he instead drew on the successful lesson he had learned from arguing for a new observatory. When Quetelet had presented the case to William, Falck and Van Ewyck, he had always asked for patience. When questioned on the possible discoveries the observatory could predict, he had replied that it was too early to tell; if he knew what the discoveries would be, he would not need the observatory.[73] In the case of *physique sociale*, Quetelet similarly defended his project on the grounds that any discussion of the absurd assumptions of his theory needed to be compared in historical context to other great movements in science, which all seemed equally absurd in their infancy. Quetelet freely admitted that his science was not 'presently at that state' to assign quantitative numbers to courage and genius, but believed it would 'one day be elevated' to that level. After all, the great model science of astronomy had one day been subject to similar charges of 'absurdity'. As he wrote, 'One cannot in effect demand from those who occupy themselves with *mécanique sociale* more than those who had foreseen the possibility to create a *mécanique céleste* during a time where there were only defective astronomical observations'.[74] Like astronomy, Quetelet argued that *physique sociale* required advances in techniques and institutions prior to advances in theory.

It would be tempting to accuse Quetelet of circular logic here. The very thing he believed true (a quantitative account of society) could only be tested once an entire set of observational institutions had been created. Yet the only justification for developing those institutions was the entirely speculative result of their construction. It was an argument that could be used for *any* project, no matter how flawed: build the institution and the theories will come. Yet Quetelet was correct that this was precisely the argument often enacted in the natural sciences, and the discoveries he had already made in physical characteristics offered reasons for promise. And in retrospect, Quetelet certainly did make good on his aim to eventually measure moral and intellectual traits. Nearly two centuries after Quetelet's work, measuring an individual's intelligence and courage no longer seem so absurd.[75]

The patience paid off. By the time of the second edition of *Sur l'homme* in 1869, retitled *Physique sociale ou essai sur le développement de l'homme*, Quetelet was able to offer a more compelling defence of quantitative moral and intellectual measures than patience. In the introduction to the section on moral and intellectual traits, he declined to repeat his earlier admission that it might be 'absurd' to statistically measure 'genius' or 'courage', only allowing that it was exaggerated.[76] In fact he was now arguing almost the opposite. Where he had previously expressed concern for the 'absurdity' of his project, Quetelet maintained in *Physique sociale* that 'one should not be surprised to hear that such a man is two times more courageous than another or three times less genius'. It was only a matter of finding the expression of behaviour that could best measure courage. After all, people often spoke of one man as being more courageous than another, and Quetelet argued that all traits 'could be expressed mathematically by saying that their courage was positive, zero, or negative'. The only difference was that these 'assessments' of courage were not precise.[77]

But could they ever be precise? Could one really measure courage in a way that would yield reliable quantitative data similar to the data for height? For Quetelet, the answer was now 'yes', and he offered a short hypothetical to answer potential critics. He first asked his readers to imagine that there were two different men who were in the same position and circumstances 'to make courageous acts' and that after one year, it was determined that one man had made 500 courageous acts and the other 300. Could it not be said then, that 'these two individuals had courage in a ratio (*rapport*) of 500 to 300, or 5 to 3'? Such a hypothetical could be extended to any number of moral and intellectual attributes, proving that a quantitative ratio was not, prima facie, absurd. Of course Quetelet understood that, practically speaking, his hypothetical was still absurd because of the 'impossibility … to place two men in positions equally favourable to acts of courage' or to accurately count such acts.[78] But this was an objection he was prepared for.

By locating the 'impossibility' and 'absurdity' of his project in the practical rather than theoretical realm, Quetelet provided the final justification for his argument. While governments may not have the ability to count *courageous* acts of *individuals*, he claimed, they do have the ability to count *criminal* acts by *groups*, so instead of counting one individual's good acts 500 times, governments could count bad acts committed by their populations 500 times. For example, Quetelet looked at the criminal conviction rates of men ages twenty-one to twenty-five and thirty-five to forty in France and was able to find that 'the penchant for crime in France is close to 5 to 3', meaning that for every 500 criminal acts committed by young men, the older age group committed 300.[79] In other words, the average man of one age group had a propensity for crime to the average man of another age group in a ratio of five to three. Quetelet was close to arguing that any quality of man, whether physical, moral or intellectual, could be determined by counting the expression of that physical, moral or intellectual act.

Quetelet argued that the reason moral and intellectual statistics seemed absurd was a practical matter of not having enough information rather than something inherently absurd about measuring morality or intellectual ability. The objection to measuring courage quantitatively could only be that government institutions do not count courageous acts as they count criminal acts, a matter that could be rectified through better means of counting and observing. Notably, this crucial foundation of the science of social physics did not rest on methodology, theory or experiment. As he wrote at the end of his introduction to book four, 'I think ... that one feels the impossibility for the moment of employing numbers in these evaluations holds rather to the insufficiency of data rather than to the inexactitude of methods'.[80] The solution was from Quetelet's point of view brilliant. Not only did it present an answer to a fundamental objection to his project – moral acts cannot be counted – but the solution of providing large amounts of data was one for which Quetelet and his network of observers were perfectly suited. The only thing standing between quantifying individual moral and intellectual actions was the practical means to gather the data. All his revolutionary science needed was a trained group of observers to do the counting.

Conclusion

Looking at the data that made up Quetelet's post-revolutionary papers, it becomes apparent that the content of the available data influenced his thinking on social physics, especially the attention to the average. Quetelet's metaphor of an average around which all other data tended to conform simply made no sense if applied to the pre-1830 data. To say that the average age of a population is (for example) 35.2 years, or that the average year of death occurred at 13.2 years of age, says little about deviation from a norm. Pictured on a graph, the average age of a population or the average year of death would not appear like a 'bell' nor are the distributions from average age Gaussian. With height, weight and some criminal data however, the average became more pronounced, because the most common 'events' tended to cluster around a centre with the endpoints indicating deviance. While Quetelet's expectations clearly affected his approach to the data, it was unlikely that the pre-revolutionary data would have led to an idea of an 'average man'.

To see the difference in the two kinds of data sets Quetelet investigated, it might be helpful to return to the odd quotation from the beginning of this study about the 'monstrous' deviations from the average. In *Sur l'homme*, he had announced that:

> If the average man were perfectly determined, one could ... consider it as a kind of beauty. All that deviated (*s'éloignerait*) from what it resembled ... would constitute deformity and disease. That which did not resemble it ... would be monstrous.[81]

Such a statement, strange as it was, simply could not have been made about the data Quetelet had been looking at prior to 1831. One could imagine, for example, in Quetelet's view an average society where all male adult citizens were 1.6 metres tall and all other heights deviations, but what to think of a society where everyone was thirty-five? Quetelet's language in the pre-1830 writings gives no sense that a four-year-old is in the same way a deviation from the norm as a four-foot tall adult. Even in the earlier writings on crime, there is no mention of averages because the data was simply not amenable to such manipulation. If the average Belgian citizen committed, say, one crime per year, there is nothing to suggest in anything Quetelet wrote before 1830 that the individual who committed no crimes per year was 'monstrous' or a deviation from the utopian average citizen who committed one crime per year. In developing the average man as a 'kind of beauty', Quetelet was almost certainly responding to the language of the data as much as the political events of the day. Much as Quetelet claimed in his earliest writings, *physique sociale* would not be carried along by speculation. It would be determined by what the numbers revealed.

This sketch of the development of *physique sociale* also shows that Quetelet designed his proposed science to the specifications of his own strength: creating large networks of data collectors. As Bourdieu imagined science to operate, Quetelet had again 'set up the standard most favorable to [his] personal and institutional capacit[y]'. While the great *savants* had moved science forward in the age of Enlightenment, Quetelet believed the next great leap in the moral and intellectual sciences of man would be made by a new kind of scientific researcher. Quetelet had positioned social physics as a science that lacked only one thing: a large group of dedicated men of science to count all of the physical, moral and intellectual traits of mankind. Such a project was an unprecedented approach to social thought, and one that provoked more than a few protests. At the time, however, few realized that while Quetelet was creating the statistical and abstract *physique sociale*, the practical demands of his science entailed such a large influx of new science workers. Distracted by debates about determinism and free will, Quetelet's critics and commentators missed this important legacy of social physics. Quetelet had not just spent the 1820s and 1830s developing a theoretical *l'homme moyen*; he had been busy developing the real average men called for by the new science of man.

6 THE OTHER AVERAGE MAN: *L'HOMME MOYEN* AND ITS CRITICS

> Every year, they say, a certain percentage has to go ... somewhere ... to the devil, it must be, so as to freshen up the rest and not interfere with them. A percentage! Nice little words they have really: so reassuring, so scientific. A certain percentage they say, meaning there's nothing to worry about. Now, if it was some other word ... well, then maybe it would be so worrisome.
>
> F. Dostoevsky[1]

A Troublesome Legacy

While social physics has been acknowledged for its direct influence on Galton, Pearson, Lombroso and Durkheim and its indirect influence on Darwin, Comte and Maxwell, it has just as often been treated as an absurd overextension of a quantifying fetish. At its worst critics derided a philosophy of crude materialism and determinism. At best, more sympathetic readings saw *l'homme moyen* as an unfortunate diversion on the road to the more mature social sciences of the late nineteenth-century. While Quetelet may have very well been that 'man of genius' called for by Helvétius in the previous chapter, *physique sociale* was just as likely to be received with something like Dostoevsky's scorn above.[2] Like the legacy of much of Enlightenment thought, Quetelet's plan for a science of man was contested throughout the nineteenth century. Understanding how *physique sociale* affected human freedom and dignity is indeed essential to understanding Quetelet's contributions to nineteenth-century scientific thought and practice, but not always in the way that Quetelet's contemporaries and commentators believed. Social physics *was* a possible new tool for controlling and defining the social body, especially the powerful idea of an 'average man' against which all others were judged, but the limits of its reach did not end at those populations under observation or those condemned to life as a 'percentage'. There was another average man to consider.

To see why the average men who practised social physics have often been overlooked, this chapter begins with the initial debates over determinism that greeted Quetelet's *Sur l'homme*. Quetelet, never philosophically inclined to begin with, struggled and often failed to deflect these criticisms. In part the charges were due to the excesses of those whom Quetelet inspired, in particular the historian Henry Thomas Buckle and his controversial work *History of Civilization in England*, which posited a strict form of historical determinism with no room for free will.[3] More surprisingly, Quetelet's biographers also contributed to this legacy, joining with Émile Durkheim and other hopeful social scientists to write Quetelet out of the history of the social sciences based in part on a deterministic reading of *physique sociale*.[4] Yet the legacy of Buckle's version of Quetelet, reinforced by a century of works on the Belgian, has led to an undue focus on the philosophical questions of free will and determinism in social physics and have obscured the more practical consequences (for both Quetelet and ourselves) of a statistical science of man. Whether counting the actions of individuals and then assigning various probabilities to their actions based on 'laws' denies the ability of humanity to freely choose is, ultimately, an unanswerable question, at least for the historian. Yet the question of what such a science might mean for those who practised versions of social physics, including institutional workers today invested in the activity of counting and predicting the actions of human beings, might be more amenable to a historical evaluation.

While the implications of social physics for its practitioners in the nineteenth century were largely overshadowed by the free will debate, this chapter concludes with a rare exception: the commentary of Pierre de Decker, Quetelet's friend, fellow academician, philosopher and future prime minister of Belgium, who may be the most perceptive commenter on the implications of Quetelet's social physics. As an academy member and friend of Quetelet, who moved in many of the networks of *savants* and statesmen, de Decker may have known better than any of Quetelet's critics what social physics entailed for the new men of science that Quetelet had been developing.

Social Physics and Free Will

Even before the publication of *Sur l'homme*, Quetelet knew his findings would be controversial. While the speculation of moral laws grounded in statistical research prompted charges of determinism, his initial investigation of intellectual qualities sparked concern that Quetelet was challenging the domain of writers and artists to express the character of a nation. Such charges began almost immediately in a particularly harsh review of the 1831 work *Recherches sur le penchant au crime*, which forced Quetelet to clarify his position before it had even been formulated. An anonymous writer had claimed that even the tentative step of determining averages was misplaced. After acknowledging Quetelet's significant progress, the reviewer claimed that 'the average man would be useless' in 'building the science' of man. The problem with an average was that it was not the kind of information that could inspire real action and reform, the practical goal of Quetelet's proposed science. After all, even in natural phenomena, average values were rarely helpful in deciding a course of action. The author pointed out, for example, that 'general topographies and average slopes are useless when trying to plan a route or direct a military operation', and that a farmer needed to know more than the average temperature of a region when deciding which crops to plant. Such was the difficulty of accepting Quetelet's founding premise that the critic called the opening paragraph of *Recherches* 'the most difficult part of the memoir'.[5]

Worse, the review also suggested that *physique sociale* would be worthless for investigations in the fine arts, one of Quetelet's most cherished hopes. While admitting that a scientific methodology was ideal for investigating nature – 'the only one that must be followed' – the arts were something else entirely. Quetelet's idealistic averages were dismissed as 'those of a friend of truth, but nothing more', and the author cautioned artists and writers that they 'need not listen to anything nor search further than the *français moyen*' to represent national character. After all, artists did not look at averages to determine the morality of the nation but instead 'exaggerated' exceptional and 'salient traits' which were 'strongly pronounced' in society. The average revealed 'nothing of the *pittoresque*' and was therefore 'best left to the sciences'.[6] The abstract average man, the foundation of social physics, looked dead on arrival.

Quetelet was stung by the review and responded at length to the criticisms in the introduction to *Recherches sur le poids* (1831), a ten-page rejoinder that was reprinted almost exactly in *Sur l'homme*. In responding to the attacks, Quetelet offered that he was not trying to challenge the ability of writers and artists to capture 'national characteristics', and that *physique sociale* did not reduce the great variation in a nation to a single average man. Dismissing the idea that his plan would miss all that is exceptional in a country, he wrote that he did not 'pretend to give the same traits, the same inclinations, and the same passions to all individuals'. Instead, the plan was to 'research the elements which predominate in certain people or of a certain age: if it is fanaticism, for example, or piety or irreligion, the spirit of servility, independence or anarchy'. Quetelet did not believe this was such an audacious claim, and that 'general' conclusions were made all the time about moral traits that 'predominated' in a given group. For example, few observers would doubt the claim that, as a group, 'one is more brave at [age] 20 than 60, or more prudent at 60 than 20'. The problem was not in generalizing, but that the generalizations were based around 'vague data'.[7] By determining the qualities in different age groups and nations with great 'precision', Quetelet believed he could assist intellectuals in understanding how moral and intellectual traits developed in different groups. With all of the data acquired so far, Quetelet asked if his sceptical readers would 'dare to affirm nevertheless that this research is useless for philosophers and *gens de lettres*?' While the science was in its infancy, he promised more to come. 'When the numeric determinations become available', Quetelet wrote, they would 'be of aid to the friend of truth'.[8]

As ambitious as it was to propose that arts and literature be founded at least partially on a quantitative account of talent and character, Quetelet tempered enthusiasm with his usual qualification about patience. He also conceded that a quantitative methodology was not enough, a concession about the limits of quantitative reasoning which was often overlooked by his critics. 'I am far from pretending', he wrote, 'that the profound knowledge of the different faculties of man is sufficient for success in the fine arts and letters'. Instead of being the only guide to arts and literature, Quetelet compared his average man to an artistic prototype, like a 'Greek sculpture', which captured an essence of the form. A modern artist could certainly look to it for inspiration, but if they were to stop there, without accounting for how mankind has changed, the resulting work 'would be cold and without effect on the spectator'.[9] Similarly he suggested that one could not stop once the average had been determined for a particular nation or group. As he insisted, but which so many of his critics misunderstood, the average man was the beginning of *physique sociale*, not the end.

The beginning was quite a bit, however, and *Sur l'homme* (1835), as well as the revised follow-up edition, *Physique sociale* (1869), gave plenty of evidence for those who saw a deterministic science at work. In order to understand the critics' attacks, however, it is necessary to understand some of what Quetelet said in these large and data-filled books. Most important is to understand that physical, moral and intellectual traits were not just analogous in Quetelet's mind because they could be quantified, but also because they changed and developed over time. While Quetelet's rhetoric and analogies to physical laws often implied that traits were static, the actual empirical data mostly described how moral and intellectual traits changed over the course of an average life. In other words, like population data, the laws suggested that what was constant was the rate of change. The full title of the work after all was *Sur l'homme et le développement des facultés*. As we will see, it was Quetelet's notion of *development* – rather than a static and unchanging social world – that was so important for Buckle's work and brought so much criticism to Quetelet's science of man.

Quetelet's link among physical, moral and intellectual development can perhaps best be seen in his discussion of English and French playwrights. Based on data from just one compendium of classical theatre in each country, Quetelet examined how the faculty of 'writing' developed over time. To do so, he listed the number of 'principle works' and their authors in age groups of five years. The table showed that in both France and England, almost no one produced a serious work prior to age twenty, and that the 'peak' years of publishing occurred between thirty and fifty, which afterwards fell precipitously. In France, for example, only five *ouvrages principaux* had been written by those at the extreme ends of the age spectrum – ages twenty to twenty-five and fifty-five to sixty – while all the other five-year age groups from thirty to fifty had produced between twenty-five and twenty-eight works. When this was plotted graphically, Quetelet's curve would once again appear. Even more intriguing was that when the French plays were arranged by 'merit' in three 'orders', the same consistency applied, even within a very small data set.[10] For example, the same age groups saw peaks in 'second order' and 'first order' works and the numbers could similarly be plotted as a curve. Just like height, the creative faculty seemed to 'grow' as one progressed in age, reaching a sustained peak around age forty before once again decreasing in frequency.[11]

More interesting for Quetelet than how 'dramatic talent' mirrored height was how well the data tracked with what he called 'passions'. Quetelet reasoned that 'passion' was not only part of the creative process, but was also the cause of many crimes, which tended to spike in the years just prior to artistic ability. This led him to conclude that dramatic talent was not a singular trait like height, but a combination of passion and 'several other faculties like imagination and reason'. He suggested that if one were to determine the growth of the faculties of passion, intelligence and reason, along with several other traits, they would find their best mixture at the ages where dramatic talent was at its highest. As seen a chart later included in the English translation for *Sur l'homme* (see Figure 6.1), the propensity for crime grew at nearly the same rate as 'literary ability'. As Quetelet wrote, 'Our intellectual faculties are born, grow, and decline, and each attains its maximum energy around a certain period of life'. Determining where, when and how these intellectual faculties converged to produce more complex traits like 'dramatic talent' was Quetelet's next step, though he warned it would take 'an infinite amount of care [and] numerous studies' to complete.[12]

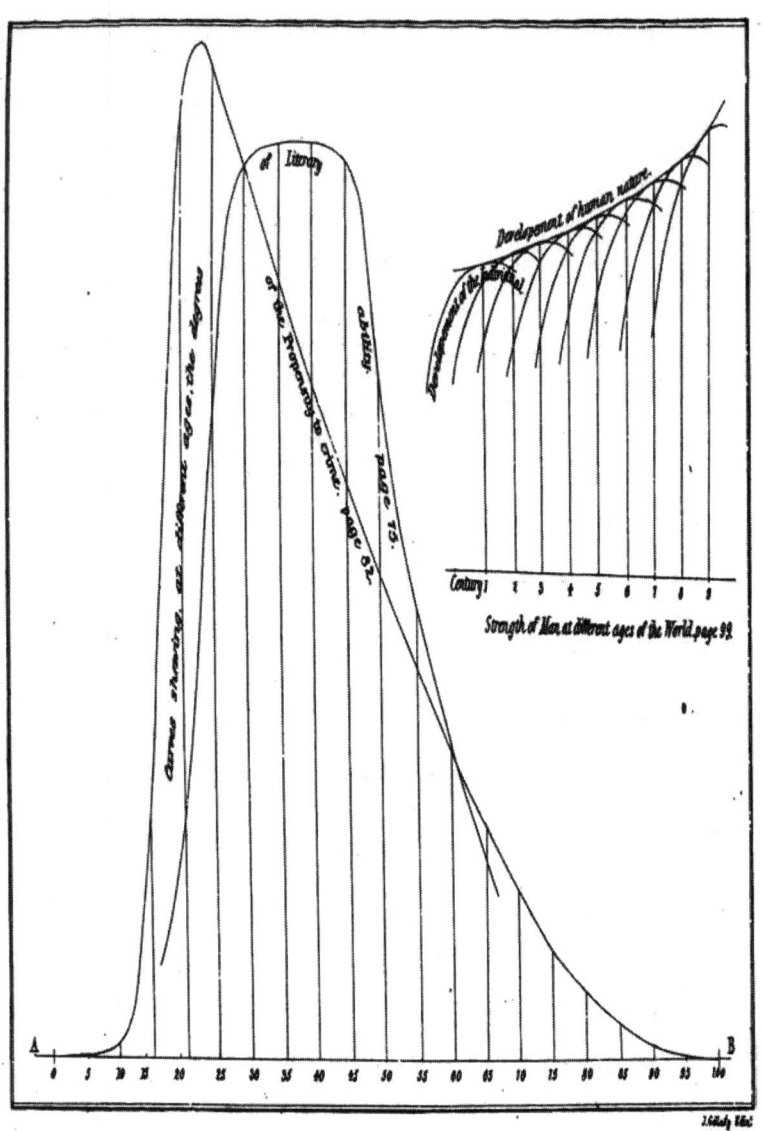

Figure 6.1: Propensity for crime and literary ability. One of Quetelet's more bold claims was that there was a relation between the 'passions' that caused a person to create art and that which led a person to commit criminal acts. In the chart shown here Quetelet shows on the left how closely the 'propensity' for literary ability matches the propensity to commit a crime. A. Quetelet, *A Treatise on Man and the Development of his Faculties. English Translation* (Edinburgh: Chambers, 1842), plate 4. University of Wisconsin Memorial Library.

While the available data on intellectual superiority (dramatic talent) was limited, with Quetelet forced to infer genius from the critical consensus on published works, there was far more available data on intellectual *inferiority*, or what Quetelet classified as *alienation mentale*. Drawing on the works of Jean-Étienne Esquirol in France, Quetelet searched to find correlations between mental illness and various 'influences'. His first conclusion supported Esquirol's assumption that there were two types of mental illness, 'idiocy, which is dependent on ... material influences' and 'folly ... which is produced by society and intellectual and moral influences'. Such a distinction was important because of Esquirol's belief that civilization itself had brought on the latter malady. Quetelet's data confirmed both points. On idiocy, after observing that Scotland and Norway had the highest rates of *aliénés*, while Wales, England and France had the lowest, Quetelet concluded that 'in the mountains there are many more idiots than in the plains'.[13] Yet 'civilization' also played a role, as the number of those considered mentally ill increased with age, or in Quetelet's language, 'exposure' to civilization.

In the same section, Quetelet investigated the age data from the Charenton asylum in Charenton-Saint-Maurice, France and found two interesting facts: the distribution by age of criminal acts seemed to match that of the distribution of 'dramatic talent', and both dramatic talent and *aliénation* increased after physical and criminal traits had reached their limits. In his words:

> The intellectual life of man and the maladies of his spirit commence to develop around the age of 25, an age when physical development has almost stopped. At this age, man has in effect almost entirely developed concerning height, weight, and strength, and it is at this limit where he has the maximum penchant for crime ... And so the average man between 25 and 30 has ended his physical development at the same age that his intellectual life has developed with more energy.[14]

The simultaneous increase in dramatic talent and diminution of physical traits again suggested to Quetelet that there was some underlying law that directed the development of physical and intellectual traits. The consistency of the data in his view prohibited the possibility of mere coincidence. Furthermore, madness also seemed to occur most often between the ages of thirty and fifty, the same years when the best dramatic work was produced. Such data might suggest to a modern reader that *aliénation* would be beneficial to playwriting, but Quetelet refused to see any connection between madness and artistic greatness. Because he had already defined dramatic talent as 'positive intelligence' and *aliénation* as 'negative intelligence', he could not make this connection, calling the simultaneous growth in artistic talent and madness merely a 'contrast'.[15]

Having examined both extremes of intellectual ability, Quetelet proceeded to moral abilities, where he was able to offer considerably more evidence for patterns in human behaviour. The reason so much data existed was because of criminal records, which were in greater abundance and more uniform than almost any other data set. Like his work on intellectual faculties, his choice of study was limited by what governments had previously decided to count. Quetelet's writings on criminality are now among his most noted legacies, and his discussion of crime has been covered in excellent detail by relevant specialists.[16] At the time, however, Quetelet's interest in crime was driven far more by the availability of criminal statistics than by an interest in understanding the criminal mind. The data was central to *physique sociale* for two reasons. Most importantly was the simple fact that, next to marriage and birth records, there was no better source of organized data for human willed activity. Not only was the number of prosecuted crimes listed for almost every department of Belgium, but the type and means of crime were listed as well. The data even tended to be recorded in comparable categories across regions and countries. Unlike judging madness or 'dramatic talent', all nations tended to have a uniform system of justice so that acts like murder and theft could be compared across borders without concern for different nomenclature or reporting techniques. Criminal data was also relevant, however, because the prosecution of crimes was explicitly tied to understanding the will to act. After all, Quetelet wrote, 'crime is not just in the action, but in the intention of those who commit it'.[17] Therefore, if a consistent pattern of activity could be found in the willed action of individual actors, then the laws of *physique sociale* would have penetrated to the most formidable source of chance and error in social acts: free will.

Quetelet began his discussion of crime with suicide data, an act closely tied to individual will in the public mind. After displaying information on national suicide rates and those of selected cities from French and German sources, he presented a table on suicide rates in Paris from 1817 to 1825, a table extraordinary in its consistency. In a population of 860,000, the number of suicides ranged only from a yearly low of 317 to a high of 396, with most years falling between 325 and 371. If that consistency were not enough, the *means* through which Parisian citizens took their lives was even more remarkable. The number of those who killed themselves with firearms ranged only between forty-six and sixty, the number of people who leaped to death from thirty-nine to forty-seven, poisonings from thirteen to twenty-eight, with similar regularity in drownings, self-asphyxiations, and stabbings. Even when compared to Switzerland the numbers were the same, with 37 per cent of Parisians drowning themselves compared to 38 per cent of Genevan residents during the same time period.[18] Following the table Quetelet remarked that there was a 'frightening concordance among the results of different years'.[19] Could the fact that about twenty people poisoned themselves every year for nine years in the entire region of Paris really be explained by the simple collection of arbitrary wills from those 180 people?

There had been regularity in human behaviour before, but what made suicide so interesting was that it was an act 'so intimately placed in the will of man'.

The meaning of suicide also varied across histories and cultures – it could either 'inspire horror' or have 'a virtuous character' – and for Quetelet it was unique in that it depended on both the external context and the internal will of the actor.[20] It was an act that was inseparable from human decision making, with an infinite number of local causes depending on the person and the means they used to kill themselves. And yet, despite all of the variables, there was consistency across time and place. The Belgian suicide data even showed that suicides increased by age group until the age group of fifty to sixty, where it again began to decline.[21] It was another curve! So impressed was Quetelet that he concluded this section with some of his strongest arguments in favour of social laws.

> That which proceeds shows us that man, in general, proceeds with the greatest regularity in all his actions. Whether he marries, reproduces, or kills himself... he always seems to act under the influence of causes which are determined and placed outside of his free will.[22]

The quote above shows the difficulty in assigning Quetelet a position on determinism or fatalism.[23] On the one hand, the case for determinism is easily made. A man who *always* acts 'under the influence of causes which are determined and placed outside of his free will' would seem a perfect Laplacian being, bound at all times in his actions by the infinite constellations of events that preceded him. We might never know all of these precedent causes, but they surely existed. Quetelet had denied that *le hasard* existed in his previous work, and though he never explicitly mentioned Laplace's demon, it would seem that his system of laws would leave no room for free actions.

Another reading of the above passage, however, leaves room for doubt over Quetelet's determinism. In the first place, Quetelet actually uses the phrase 'free will (*libre arbitre*)', which is not mentioned in Laplace's *Essai philosophique sur la probabilité*, indicating that Quetelet at least recognized the concept. If something could be 'outside free will', this presumably meant that some things were covered by free will as well. Secondly, the final sentence refers to 'man in general' rather than an individual. Most who criticized determinism were not focused on 'man in general', but whether *individuals* had free will. As Quetelet had already made clear when discussing the law of large numbers, free will was effaced when one combines the free actions of the many; it is not absent before the numbers are combined. Finally, the conclusion was provisional, as Quetelet claimed that man only 'seems to act' under these laws. The nature of the laws themselves was unknown and, as far as could be determined, unimportant. Free will within the individual was not denied by *physique sociale*; it simply was not covered by the new science. Such apparently contradictory passages occur throughout Quetelet's entire body of work, which in large part explain the controversy around *Sur l'homme*.[24] Introductions to his works in the 1820s and 1830s began with praise for the law-like regularity of social behaviour, the inevitability of crime and the absence of the individual will at the level of large numbers. Immediately after describing this apparently deterministic science, however, Quetelet offered prescriptions for governments to alleviate human suffering and the belief in the perfectibility of mankind. This was the concern about the 'budget of the scaffold, prisons, and chains' that had motivated his earliest work. Such nuances (or outright contradictions) did little to alleviate his harshest critics.

While Quetelet vigorously defended himself against the determinist label, it was likely that the objections were more an annoyance than a philosophical matter to be seriously investigated. Debates over free will in fact played almost no role in the first decade of his statistical research and were rarely mentioned except for a brief two-year period in the 1840s. For example, the 1835 edition of *Sur l'homme* mentions '*libre arbitre*' only once in over 700 pages and then in reference to a comment by Villermé.[25] By comparison, however, the 1869 version contains no fewer than fifteen references, all in sections added to the original that drew on two papers published in 1847 and 1848. Still the philosophy was buried deep within the endless tables. Reading Quetelet's defence of *physique sociale* in the context of his overall writing makes clear that while Quetelet was struck by these charges, his most pressing concern was for the matter to be dropped so that the project of data collection could be continued. Provoking philosophical arguments was never the plan of social physics.[26]

Quetelet als Queteletismus: Buckle and the Biographers

Despite Quetelet's lack of interest in the subject, or perhaps because of it, the connections between *physique sociale* and free will endured. Though few studies employ the term *Queteletismus* – a mocking description of Quetelet never used in French or English commentary – Quetelet and *physique sociale* have become synonymous with a long discredited theory of vulgar statistical determinism.[27] While Quetelet played little role in a philosophical debate proper, historical accounts of the era often treat the name 'Quetelet' as shorthand for a brief and failed pseudoscientific vision of statistics.[28] This legacy is not due to Quetelet's own interest in the philosophical problem of determinism, however, but to later followers, critics and historians. In the words of one historian, Quetelet has suffered a 'double injustice' at the hands of his critics because 'most commentators ... confine[d] themselves to theoretical and quasi-philosophical generality'.[29] While all writers must account in some sense for those they inspired, the path from Quetelet's original writings to his portrayal at the beginning of the twentieth century was marked by exaggeration and the conflation of ideas. Despite Quetelet's insistence that he did not deny free will to individuals, one enthusiastic follower and three ambitious sociologists ended up defining *physique sociale* in opposition to its creator's hopes. In their focus on abstract laws, forces and philosophical speculation, these authors set the context for a century's long misunderstanding of the consequences of social physics.

The enthusiastic follower was Henry Thomas Buckle (1821–62), who published the first volume of his 'Introduction' to the *History of Civilization in England* in 1857, two decades after *Sur l'homme*. Buckle's unfinished work (the incomplete 'Introduction' itself totalled three volumes and over 1,200 pages) was an attempt to create a true science of history based on an odd admixture of influences including Comte, Condorcet, Mill, de Maistre and Quetelet. The final work produced a thoroughly and explicitly deterministic historical study.[30] Following a generation of writers in the shadow of Condorcet and Laplace, Buckle wrote that he wanted:

> to accomplish for the history of man something equivalent ... to what has been effected by other inquirers for the different branches of natural science. In regard to nature, events apparently the most irregular and capricious have been explained and have been shown to be in accordance with certain fixed and universal laws.[31]

Buckle's belief was that human history had developed according to these 'fixed laws', with physical characteristics gradually diminishing and being replaced by moral and intellectual qualities. The development of these laws were supposed to follow the development of Quetelet's average man, though Buckle reduced the causes to what he called four 'physical agents': climate, food, soil and the 'General Aspects of Nature'.[32] The combination of these effects could explain the 'national character' of each nation, and resulted, unsurprisingly, in a declaration that Great Britain was the most mature nation. It was, in short, Whiggish history *avant la lettre*; an unapologetic defence of Britain's greatness and standing in the European world.[33]

Buckle's plan of national progression was indebted to the stadialism of Comte, but also mirrored the development of Quetelet's average man. Just as Quetelet had surmised that the average man's physical powers reached their limit shortly before the growth of their moral and intellectual traits, so too did civilizations need to maximize their physical exertions before moving on to intellectual matters. While some nations were like the twenty to twenty-five-year-old male – at the maximum of physical qualities and the minimum of moral and intellectual qualities – other countries, particularly England, had left behind their physical development and were now in the process of attaining their maximum moral and intellectual development. As Buckle wrote after a broad survey of works by Quetelet, Hegel and Mill, 'it may be fairly inferred that the advance of European civilization is characterized by a diminishing influence of physical laws, and an increasing influence of mental laws.'[34] Quetelet's average man had become the model for the Rise of the West. Perhaps the century's most confused extension of the logic of French Positivism, German Idealism and British Liberalism into historical study, Buckle's book would do more to cement Quetelet's negative legacy than any other work.

Quetelet could not have asked for a worse exponent of social physics. As one review of the *History of Civilization* phrased it, Buckle was a master of 'acquisition' and a disaster at 'construction'.[35] Characterizing Buckle as well read but incapable of original thought, the review opened with the opinion that, 'If the author of this book has failed to give us a history of civilization, it is due to his mental deficiencies, and not to those of his acquirements'.[36] The complaint against Buckle was that while he had read nearly every important work in the past half century, he had failed to understand any of what he read, Quetelet included. He had taken Quetelet's theory and applied it to history, yet the reviewer noted that 'historical method' required 'continuity of thought and clarity of mind ... which Mr. Buckle does not possess'.[37] Though Buckle's unfinished work received some positive notice and sold spectacularly, at the time an association with *History of Civilization* made Quetelet appear the most vulgar determinist.

In a work which broadly referenced a hodgepodge of theories of history and science, how was it that Quetelet stood out? In spite of the many connections made between Buckle and Quetelet by critics over the past 150 years, the Belgian was mentioned less than a half-dozen times in over 1,200 pages. Quetelet was not even given an entry in the index for the abridged 1885 edition of *History of Civilization*. Yet it has been claimed that 'Buckle ... founded his doctrine on Quetelet'.[38] Buckle's vision *did* owe some of its ideas to Quetelet, but the influence was more to do with Quetelet's stages of development than any iron laws of statistical determinism (see Figure 6.2). Quetelet's laws of development, as seen in his interest in ameliorating crime and poverty, had far more flexibility than anything Buckle imagined. Any uninterested observer who read all of *History of Civilization in England* would find it difficult to see much of anything related to the work Quetelet did on *physique sociale*.

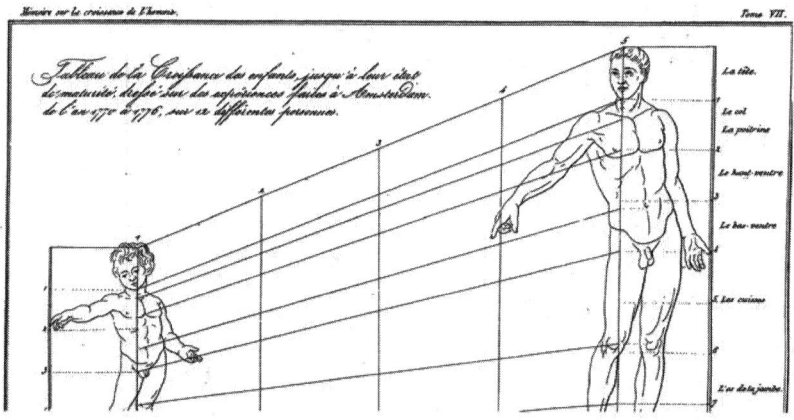

Figure 6.2: Tableau de la croissance. One of the reasons Henry Thomas Buckle was so drawn to Quetelet's ideas was that they allowed for both growth and regularity, as seen here in the physical growth of the average man. Though Buckle did appropriate Quetelet's average man as an idea of growth, Buckle's iron laws of determinism were nowhere to be found in the Belgian's social physics. A. Quetelet, 'Sur le poids de l'homme aux différens ages', *Nouveaux mémoires de l'Académie royale des sciences et belles-lettres de Bruxelles*, 7 (1832), pp. 1–44. Inset following p. 44. University of Wisconsin Memorial Library.

Quetelet's link to Buckle was rather due to later statisticians who, in applying the derogatory term *Queteletismus* to *History of Civilization in England*, were hoping to discredit an idea of statistics though an attack on its least subtle proponent. Buckle's work had gone through seven editions in Germany, mostly it seems as a sort of cautionary example for studying history. In the Idealistic tradition, German objections drew from the well of *statistik*, which had largely dismissed quantitative measures, and instead emphasized qualitative accounts of state development. As Ted Porter has put it, 'The identification of statistics with numbers ... was resisted by German academic statisticians until the 1850s and 1860s'.[39] Furthermore, as Ian Hacking noted, the belief in 'statistical laws' was anathema to 'Kant's heirs', who protested against the conflation of regularities and laws they felt typified French theorists. Even the lone enthusiast east of the Rhine – Adolph Wagner – eventually recanted his excitement for Quetelet.[40]

The German reaction first appeared in G. F. Knapp's series of critical articles on Quetelet and found its way into the mainstream through a biography of Quetelet by the Swiss author Naúm Reichesberg: *Der beruhmte Statistiker Adolf Quetelet*.[41] Both Joseph Lottin and Frank Hankins started with Reichesberg, and through this work many of the German critiques made their way west. Though in some ways appreciative of the project, Reichesberg referred to free will as 'the cavity (*leere Schall*) of [Quetelet's] theory', and many German-language critics agreed.[42] Notably, however, the complaints did not occur until almost forty years after *Sur l'homme* was released and coincided more closely with Buckle's publication in German than with Quetelet's writings. Indeed, the term *Queteletismus* does not appear to have been used prior to 1871, thirty-six years after *Sur l'homme*, but just eleven years after *History of Civilization of England* appeared in Germany.[43] While Reichesberg and the Germans had many positive things to say about Quetelet's career, the severe reaction to the determinism of Buckle have since shrouded interpretations of Quetelet.[44]

Yet there is no reason to see social physics through the lens of Buckle, as the latter was such a 'compiler' that nearly any social or historical idea of the first half of the nineteenth century could be taken up through his writings. Indeed, it is possible that the negative German response to Buckle could be conceived without invoking *Queteletismus* at all. The harsh criticism of Buckle could just as easily have been a challenge to what Michael MacLean has called the 'crude materialism of ... scientific popularizers in Germany ... [that] enjoyed considerable popularity among the German middle classes'. In fact MacLean's survey of the reception of Buckle and the tremendous historiographical debate that ensued mentions not Quetelet, but Comte, as the principal target of German attacks. The concern with Buckle was only at the level of stomping out faddish ideas rather than a serious philosophical debate.[45] In a twentieth-century introduction to an abridged version, Quetelet is not even mentioned, with Comte and Mill the primary concern.[46]

Had the harsh reception Quetelet received in Germany been localized, the overriding focus on free will may not have followed Quetelet into the twentieth century. There were certainly other critics, especially in England. John Herschel in 1850 had expressed concern as well about Quetelet's notion of free will before Buckle arrived, but only in the context of a positive review of his friend's work.[47] John Venn, too, had dismantled the determinist version of Quetelet in 1866 by associating the Belgian with Buckle's vision.[48] Yet the German-language interpretation of Quetelet became definitive because Quetelet's French and English biographers took Reichesberg's survey and Knapp as their starting point, a decision that continues to influence writing on Quetelet into the twenty-first century.[49] If the early critics' focus on free will has prejudiced later interpretations, it has been largely the result of the two biographies, along with criticism from one of the most famous social scientists at the turn of the twentieth century. Joseph Lottin, Frank Hankins and Émile Durkheim all had different reasons for accepting the verdict of Quetelet as a determinist. Durkheim, who had been lumped by Reichesberg with Quetelet and Buckle as part of the oddly named 'French school', was largely trying to clear his own name. Lottin and Hankins, who produced the most cited biographies of Quetelet, on the other hand were driven by their own concerns about creating a science free from determinism.

The influence of Quetelet on Émile Durkheim (1858–1917) is now recognized even at the level of the textbook, but the connections between *physique sociale* and Durkheimian sociology were not due to an acknowledgement of debts in the latter's writings.[50] Durkheim actively worked to distance himself from Quetelet, noting initially the Belgian's influence, but mentioning it less frequently in his later methodological works.[51] In his first major work, *Suicide*, Durkheim claimed that *l'homme moyen* was 'the only systematic explanation of the remarkable phenomenon' of stability in social data.[52] No one before had been able to explain the fact that acts apparently due to free will showed such consistency from year to year. Yet even this recognition was a kind of backhanded compliment, as Durkheim's praise for *l'homme moyen* was intended largely as a preliminary to its dismantling. Durkheim's explanation of suicide needed to replace *some* theory, and Quetelet's was the most likely target. Just two pages later, he claimed that 'Quetelet's theory rests on an inaccurate observation' because it required social forces to act on individuals at an evenly distributed rate. For an event as rare as suicide, Durkheim argued, this was an absurd proposition, because it meant that in France, for example, the average individual would have an 'inclination' towards suicide of just .015 per cent, a minuscule amount that could hardly be a factor in determining any one single suicide.[53] This was, for Durkheim, exemplary of the larger problem of *l'homme moyen*. In a passage unsupported by reference to any particular work, Durkheim then claimed that:

> [Quetelet] thought it certain that stability occurs only in the most general manifestations of human activity; but it is equally found in the sporadic manifestations which occur only at rare and isolated points of the social field. He thought he had met all the requirements by showing how, as a last resort, one could explain the invariability of what is not exceptional; but the exception itself has its own variability, inferior to none ... There are ... very few people who kill themselves; the great majority of men have no inclination to suicide. Yet the suicide-rate is even more stable than that of general mortality. The close connection which Quetelet sees between the commonness of a quality and its permanence therefore does not exist.[54]

It is hard to maintain the objection against Quetelet that he made a 'close connection' between 'common qualities' and 'permanence'. Many of the most interesting data found in *Sur l'homme* – from the articles on dramatic talent to criminal acts to suicide itself – focused on the fact that *uncommon* events showed just as much, if not more, regularity than common events. Quetelet had spent so much time on suicide precisely *because* it was 'exceptional'. This was why he became so interested by the fact that twenty Parisians poisoned themselves every year. Suicide was a focus of *physique sociale* for the same reason it intrigued Durkheim: it was a rare event that was intimately connected with the will *and* it was repeated with regularity. In fact, in *Sur l'homme*, Quetelet even anticipated Durkheim's famous speculation that modern *anomie* was a cause of suicide. Observing that suicides had increased significantly in Berlin, Quetelet posited as far back as 1835 that suicides might in fact 'follow from civilization'.[55]

Durkheim's conclusion that the average man could not explain suicide seems to be a misunderstanding of the project. For Quetelet suicide data was a justification for believing in a *physique sociale*; it was not the case that he thought a statistical law acted on each individual with equal force. Durkheim seems to have attributed his own independent social force to Quetelet's *l'homme moyen* and then complained that the average man was inadequate. Yet Quetelet never claimed *l'homme moyen* exerted any force on individuals. In the case of suicide, the average man showed little propensity to kill himself, and so therefore more specific and local causes needed to be found. In *Du système sociale*, Quetelet had even spoke of the 'growth' of a *penchant* for suicide in individuals as they became older, but this did not mean they were more influenced by the law itself, but rather by some external cause like exposure to civilization. Quetelet would have largely agreed with Durkheim that the solution for the problem of suicide was to change social institutions instead of attending to the individual.[56] Though there is no evidence that Durkheim was thinking of Buckle, it had been Buckle, not Quetelet, who stressed that research on suicide revealed the actions of social forces on individuals.[57] One of Buckle's most infamous claims, repeated often in the historical literature, was that the 'power of the larger law' of suicide '[was] so irresistible, that neither the love of life nor the fear of another world can avail anything towards even checking its operation'.[58] Quetelet however had said nothing of the sort. An 1857 reviewer of Buckle who was particularly critical of the passage above makes no mention of Quetelet in fact, instead blaming Buckle's 'hopeless and unpromising philosophies' on 'the irreligious instinct of Comte'.[59] No matter the source of the confusion, the verdict from one of the great names of sociology had been delivered.

Durkheim had good reason to want to distance himself from Quetelet and Buckle; he had been seen as part of an 'astronomist' group criticized by Reichesberg and was looking to take the field of sociology far away from the reductive certainties of Buckle. It may be more surprising however that Quetelet's two primary biographers, both of whom were largely sympathetic to the Belgian's project, also contributed to a perception of *Quetelet als Queteletismus*.[60] Published just four years apart over a century ago, the twin biographies have been responsible for an overemphasis on Quetelet's determinism problem, largely ignoring most of his methodological and institutional work in favour of disentangling the philosophical and intellectual issues. Both drew significantly from Reichesberg and therefore incorporated the German-language tradition's concern with free will. Subsequent histories of the maturation of the social sciences have not looked at Hankins and Lottin uncritically, but the influence of these books has been such that a reappraisal of the work might be necessary. It was after all only forty years after Quetelet's death when the two books were published and over one hundred years have passed since their publication.

If Quetelet had been a great philosopher, then Joseph Lottin's *Quetelet, statisticien et sociologue* (1912) would have been a classic of intellectual history. As it is, it has remained the starting point for most efforts to understand Quetelet's project to construct a science of man.[61] A professor of philosophy who had graduated with a degree in theology from University of Louvain, Lottin was an unlikely scholar to publish such a thorough history of a statistician. His earlier writings had been concerned with refuting the determinism he saw in John Stuart Mill, and after his monumental work on Quetelet, he seems to have never published again.[62] Yet his work on Quetelet covered nearly every possible intellectual angle of *l'homme moyen*, from what Quetelet meant by each of the many types of 'causes' to a one-hundred-page section on 'Free Will and Social Causes'. Inspiring a trend, Lottin directly compared *physique sociale* to four other models of 'social science': Comte's *physique sociale*, Süssmilch's *Ordre divin*, Condorcet's *Mathématique sociale* and Laplace's *Mécanique sociale*, eventually settling on the last as the most similar project. In concluding that *physique sociale* was thoroughly Laplacian, Lottin may have been using his fellow Belgian as a proxy to attack Laplace.

The absence of a coherent philosophy in *l'homme moyen* necessitated that Lottin substitute his own ideas. After reviewing the range of responses to Quetelet – from strict determinism to compatibilism to incoherence – Lottin offered fifty pages on 'The Nature of Quetelet's Determinism' in which he then declared that 'Quetelet is not a philosopher, but a statistician, mathematician and ... poet'.[63] Subsequently, however, Lottin proposed a philosophy for Quetelet, which owed as much to his own writings on free will as to Quetelet's.[64] As one review of *Quetelet* noted, 'the citation is so mingled with exposition that in some places the reader is in doubt whether the ideas presented are those of Lottin or Quetelet',[65] and that at another point Lottin's 'position is not distinguishable from that of Quetelet'.[66] *Quetelet* became a way for Lottin to work out his own interests, and he eventually settled on a reading of *physique sociale* that allowed room for individual free will within a socially determined world: 'the thesis of free will accommodates itself easily with *sociological laws*'.[67] It was a comfortable position for a theologian who also wanted to investigate social laws.

Lottin's conclusion that '*sociological laws*' were not a threat to individual wills was based more on Durkheim's understanding of social forces than Quetelet's, as the latter never used the term. As Lottin explained, sociology was 'by definition' the study only of 'collectives' and had nothing to do with 'psychology' because it 'did not directly occupy itself with individual activity'.[68] The statistician would do well in fact to 'rule out' all considerations of the individual mind. Such advice is reminiscent of Durkheim's methodological writings, where he also worried that sociology would become a 'corollary of psychology'.[69] Indeed, Lottin's analysis of *physique sociale* was often similar to Durkheim's, where a form of the latter's belief in 'social forces' was read into Quetelet's own ideas, thereby making *l'homme moyen* seem to be an overly deterministic force. Quetelet, however, never claimed that *physique sociale* dealt with individuals. Instead he only remarked that due to the large number of observations, acts of free will 'cancelled each other out' when large populations were studied. Both Lottin and Durkheim had misread *l'homme moyen* as an autonomous force rather than an ad hoc description. Where Quetelet saw only data, the sociologists had interpreted a distinct entity.

This confusion was noted by another early twentieth-century biographer, the statistician and sociologist Frank Hankins (1877–1970). Hankins, president of the American Sociological Association, who notably titled his 1908 work *Adolphe Quetelet as Statistician* (rather than Lottin's 'Statistician *and* Sociologist'), complained in a review of Lottin's biography that the Louvain philosopher 'seem[ed] to think of cause as an active force'.[70] Furthermore, the excessive focus in Lottin's work on philosophy overshadowed Quetelet's key contributions to statistics. After a string of critical comments, Hankins concluded of Lottin that 'in general, it may be said that the author shows no familiarity with the recent developments of scientific method'.[71] Hankins's critical view of Lottin's work could be attributed to the fact that he had published a competing work on Quetelet just four years prior, but it becomes clear from a parallel reading of the two works that Hankins had vastly different concerns; indeed, the American never mentioned Durkheim and only briefly touched on Quetelet's work outside of statistics. Hankins had no more interest in 'social forces' than he did in dialectical materialism.

Hankins instead focused on the implications of Quetelet's work for statisticians interested in determining population 'types'. Due to his own interest in population genetics and evolutionary statistics, Hankins went searching in Quetelet's writings for advancements that could anticipate the more mature science that developed at the turn of the century in biology. Not surprisingly, he found Quetelet's average man, developed in the late 1830s, a poor guide to biological evolution. More a student of Galton and Pearson than the German and French sociologists, Hankins avoided almost all talk of 'social forces' and 'collectives' and instead limited his attention to three areas: Quetelet's role in history in statistics proper, the value of *l'homme moyen* for statistical analysis and the possibility of doing moral statistics.

In the first two areas of focus, Hankins found Quetelet's work to be valuable, but nothing that rose to the level of a true discovery. Largely in agreement with Lottin, who believed nearly everything in Quetelet could be found in Laplace, Hankins concluded that Quetelet's 'genius consisted not so much in original conception' but in 'gather[ing] up the chief statistical tendencies of his time, and contribut[ing] ... to the advancement of each'.[72] His review of the average man fared little better, with statements of confusion at every turn. Consider a small sample of Hankins's comments on Quetelet:

> Just how Quetelet reached the conclusion that the average man represents perfection in mental and moral traits in not equally clear.
>
> What he meant by an excess of mental ability or morality or health Quetelet nowhere clearly states.[73]
>
> Quetelet was by no means consistent as to the permanence of the type.
>
> It seems impossible to reconcile [his] various propositions.[74]

In view of these concerns, it might be asked why Hankins still tried to tease out a consistent position for Quetelet's average man. Why not say it was an incoherent and forgettable concept and be done with it? The answer was that Hankins's Quetelet was that typical character of early twentieth-century histories of science: the figure grasping at discoveries just out of reach. For Hankins, Quetelet was close to no greater scientific breakthrough than discovering evolution through natural selection. As he wrote of the average man, 'it is interesting to note how very close Quetelet came to the discovery of the selective action of environment. He was only one step from it'.[75] Hankins also claimed that 'There are other passages in Quetelet which likewise suggest a selective process in nature'.[76] In summing up how Quetelet's inconsistent and unclear theory of *l'homme moyen* almost led to the most revolutionary theory in nineteenth-century science, Hankins wrote:

Thus, though he used many expressions suggestive of evolutionary change of the type, he did not grasp the notion of such change. Though he developed and used the method which has come to serve in the work of Galton, Pearson, and others as the basis for the mathematical demonstration of evolutionary development, he did not himself make any such use of it.[77]

Quetelet's critics accused him of many things, but among the most surprising may be his inability to 'make use' of a theory developed two decades after his death.

Hankins was more charitable towards Quetelet in his contributions in moral statistics. He agreed that he had a better claim to the title of the founder of sociology than Comte and that in Quetelet '[was] to be found the basis of the quantitative study of social life'.[78] More interesting perhaps was Hankins conclusion that far from *physique sociale* being the determinist minefield found in Lottin, Quetelet's work actually helped to diffuse the explosive issue of causation in moral acts. Through his many inconsistencies and variegated terminology of causes, Quetelet had stumbled upon a way to analyse *correlation* rather than causation.[79] As Hankins put it, 'From this simple and general basis given by Quetelet, the difficult problem of studying causal relationships has been advanced to a method of quantitatively measuring the correlation of two variable elements throughout their distribution'.[80] This 'advancement' may have been the single most important development for professional statistics and maybe the social sciences because it allowed the field to supplant a morally fraught term (cause) with a technical one (correlation). Even here, however, Quetelet's greatest importance could only be found by reading later discoveries back into his work.

Quetelet, of course, was no more trying to anticipate Galton and Pearson than he was Durkheim. Though differing significantly in manner and focus, both Lottin and Hankins approached *physique sociale* from the perspective of what it could say about an autonomous society. Whether it was the belief in a separate sphere of social action, the influence on individual free will or the applicability to studying natural selection on social and physical bodies, they shared the idea that *physique sociale* was at its heart a theory about something called *society*. Yet rarely did Quetelet frame social physics as a theory or a set of hypotheses, and he certainly never accorded it the influence on free will or causation that his biographers or later historians have claimed. Given the motivations of Durkheim, Lottin and Hankins, their focus on the intellectual and philosophical consequences of social physics is hardly surprising. Interested in independent forces, the philosophical problem of free will and the field of statistical biology, respectively, they took little interest in the many other aspects of *physique sociale*. In particular, they ignored the element of social physics most interesting to Quetelet: the expansive network of *savants* and researchers that would be needed to carry out the project. As discussed in the next section, it would take a colleague more familiar with Quetelet's science – a fellow member of the Brussels Academy no less – to mention the most practical implications of social physics.

The Other Average Man

By 1846 Quetelet was close to publishing two works that he believed would solidify *physique sociale* as *the* science of man and statistical probability as its proper methodology.[81] Though it would be another decade before Buckle published the bestselling account of British history based on Quetelet's writings, Quetelet had concerns over early readings of his 1835 work that emphasized the deterministic possibilities for a science of man. Therefore, as part of his presentation *Sur la statistique morale et les principes qui doivent en former la base* (1848), Quetelet invited two colleagues to comment on the question of whether social laws entailed determinism. In evaluating these comments from the people who knew him best, in particular those of the Belgian philosopher Pierre de Decker, a different vision of the relationship between social physics and free will might emerge.

The most direct argument in support of Quetelet came from the eclectic philosopher Pierre François van Meenen (1778–1852), who did Quetelet few favours in attempting to defend his friend from charges of materialism. Getting straight to the point, van Meenen declared that social physics 'as a science, does not entail materialism, fatalism, or an attack on morals and religion'. Dealing only with a large group of people, it did not ask questions of individuals. After asking if 'free will posed a problem to discovering the laws of society', van Meenen gave an unqualified 'no'.[82] In fact, if the laws were true, they must be 'providential' in origin and therefore nonthreatening to a belief in free will.[83] And in any case, Quetelet's laws of social physics were no more than mere 'limits' within which freely acting individuals could choose their path. The laws were 'elastic' and 'without violence and rigidity' allowing plenty of room for 'life, thought, and even liberty'. They were, simply, '*laws of possibility*'.[84] Summing up his position as an unqualified defence of social physics, van Meenen claimed that 'one can therefore do statistics, even moral, even individual, without placing human liberty in question'.[85]

So far, so good. Yet in his positive affirmation of statistics, van Meenen could not help chiding those who objected to such sound logic. Declaring that objections had been launched only by 'men of mediocre understanding', he claimed no 'intelligent and attentive reader' who had a 'true sense of [Quetelet]' could think that human free will could be impinged upon through statistical inquiries. Thought Quetelet rarely deployed the kind of straightforward language of the following paragraph, he may have enjoyed hearing van Meenen's incredulity at objections to social physics.

> How strange! Here are facts submitted to calculation, of which the results more or less surprise us. The facts are constant, the calculus exact, the results rigorously deduced. What to do? Applaud a new discovery. But this new discovery is contrary to our systems! So what if this is true? Can our systems not be in error? Allow therefore the author to continue his studies, and we, we re-examine our systems.[86]

Quetelet agreed that it was 'absurd' to suppose social physics could inhibit the free will of a single individual. Although conceding that 'the free will of man fades' in statistical calculation, this was only due to the high number of observations. Only at the level of administrators or statisticians did free will not enter into the picture.[87] It was to be assumed from these remarks that the people who conducted social physics, for whom free will was not a question, were the 'intelligent and attentive' readers for whom van Meenen had called. These were most likely also the researchers and academicians Quetelet and van Meenen had supported in the universities and institutions of science in Belgium.

In presenting statistics as a means to observe general laws, both van Meenen and Quetelet made explicit reference to the fact that the question of free will was dependent upon the vantage point of the observer. Yet the third speaker, Pierre de Decker (1812–91), made a different case for when 'free will fades', looking not at the *individual under observation*, as Quetelet and van Meenen had done, but at the free will of *those making observations*.

For the Catholic de Decker, the question of *libre arbitre* was obvious: 'all proclaim' that 'man is free', and 'from inside and all around us the consciousness of his freedom provides his strength, his dignity, and his grandeur'.[88] Yet such freedom was bounded, and de Decker explained to the audience that, from Plutarch to de Maistre, it has been recognized that mankind is connected through a 'supple chain' to a larger order and stability.[89] God of course provided one level of the chain, but de Decker also explained that free will is in fact limited at all levels of man's life, and that free will did not make 'laws', 'contracts', 'Kings' or 'nations'.[90] Contrary to the 'unjust pretention' that a 'social contract' entails the voluntary loss of will, de Decker quoted de Maistre and Henri Fonfrède to the effect that such laws of human society were outside of the free actions of mankind. Though conservative counter-revolutionists may have made strange support for Quetelet's enlightened social theory, de Decker provided a defence (and understanding) of social physics that was rare for the era.

Physique sociale, in fact, was for de Decker a clear recognition that mankind has severe limits on its collective freedom of action. While he claimed that an individual is 'free and responsible' at the level of '*personality*', his or her freedom 'disappears' in the '*social milieu* where one lives'. No wonder Quetelet's social physics effaced free will – it studied acts of collective societies, states and nations. Quoting Fonfrède, de Decker pointed out that 'society is not produced by the reasoning or will of man – it is the *necessary* result of the elements from which God has composed human nature'.[91] Even in individual reforms and initiatives de Decker believed that man has 'exaggerated singularly the role which it is called to play'.[92] In sum, Quetelet's quantitative studies served as a reminder that 'the human mind must know to respect the limits assigned to its actions'.[93]

Though de Decker was a philosopher at Louvain at the time, his defence of Quetelet's statistical calculation was not a simple rhetorical move, or an academic exercise in the determinism debate. His instinct to 'limit' human actions had a strong practical basis, especially in his later career as prime minister. Though he was no fan of the United Kingdom of the Netherlands (he was once called 'very religious, very Catholic, and very anti-Dutch'), de Decker did seek to retain some level of order after the chaotic years of his youth, and it seems likely that in limiting 'free will', he saw *physique social* as a conservative check on revolutionary fervour.[94] In particular, he noted that the people who conducted social physics and those who ran the country had much in common. The men that Quetelet had been creating in Belgian institutions of science would be ideal for both science and state. Echoing Quetelet's language when he described the 'rare men' of the Académie royale, de Decker explained that those who 'carr[y] *personal, spontaneous, and humaine* elements in to the institution' were not 'called to rule ... society'. Rather, it was those who gave up their individual elements, as well as their free will, who had the capacity to 'dominate society'. More pointedly, de Decker quoted Quetelet that '*the important role of moral statistics is to show to the legislator the point where he must act to modify the social state*'.[95]

De Decker had somehow, in the course of an ostensible discussion of free will, come to describe the 'rare' and 'new' men of science that Quetelet had been developing in Belgium for the past twenty years. The descriptions were of the kind of science and state careers that Quetelet's friends in Ghent has objected to and harkened back to the sacrifices made for the divisions between the sciences and literature in Belgian education. De Decker's words also described the new kinds of workers who staffed observatories and other large institutions of science, and they were nearly identical to the criteria that Quetelet had given for inclusion into his history of Belgian *savants*. While trying to explain why social physics should not be considered materialist or determinist, de Decker had managed to explain perfectly Quetelet's practical project to create the institutions of *physique sociale*.

De Decker's argument for his friend's science seemed to lead to an ironic conclusion. Quantitative statistics did not alter or impinge upon the free will of people who were being studied or made subject to observation, the complaint made by Dostoevsky in *Crime and Punishment*, mocked in the person of Gradgrind in Charles Dickens's *Hard Times* and levelled to this day against many institutions of observation. According to de Decker, *this* free will was inalienable, given by God and forever located in mankind no matter what the numbers said. *Queteletismus* was simply a mistaken impression of what free will was; no institution of man could ever take away a gift from God. Yet free will certainly *was* lost in the creation of social physics in de Decker's reading, only not by those *under* observation but by those *doing the observing*. Like those members 'called to rule' in state-sponsored institutions of science, the practitioners of social observation needed to give up their 'personality' and 'humaneness' in order to 'dominate society'. Quetelet had already made as much clear in his praise for Belgian men of science, his cultivation of sources for the *Correspondance* and his examinations of observatories in France and Germany. The old world of eclectic men of science had clearly needed to be replaced by a group of more common men in the pursuit of large-scale projects. Though Quetelet called them 'rare' men or 'new' men, they might also, in their limitations of individual 'personality' and necessary commonality, be considered average men.

Conclusion

Because of Buckle's brief fame, the methodological concerns of early twentieth-century sociologists and the subsequent historical work based on these sources, the determinist consequences for *physique sociale* may have been exaggerated. Even though Quetelet did address the question at times, the consequences for free will were far from the focus or even a primary aspect of his overall project. Quetelet's project was not Buckle's. Considering that Quetelet had ten times as many papers on shooting stars as he did on 'laws' and that the data sets and tables far surpassed his occasional nods to philosophy, it is hard to reconcile much of the historical literature on *physique sociale* with the proposed science itself.

Even the earliest German critics, many of whom were academics or state employees, who mocked the idea of a deterministic *Queteletismus*, may have been as concerned about the practical consequences of losing their own free will as they were about the abstract conclusions of the science. As Porter has written, in Germany, 'numbers were generally identified until 1860 with the official form of statistics, a "lifeless and mechanical" practice that flattened the delicate social contours ... beneath the homogenizing force of bureaucratic centralization'.[96] They might have suspected they were being made into average men. This was de Decker's insight as well, yet he did not have nearly the same concerns: such was the nature of bureaucracies and states that some freedoms may be lost. Interestingly, while de Decker grounded his idea of social physics as state service in the conservative, anti-Revolutionary spirit of writers like de Maistre, this has rarely been the political interpretation of social physics or much of the post-Enlightenment sciences of man, from the nineteenth century until today. Hacking has claimed that the Germans initially rejected the 'probability-monger' Quetelet because his 'liberal' and 'utilitarian' statistics 'had no appeal in the wholesale, conservative view of society'.[97] But in the language of de Decker, de Maistre and Fonfrède, it seemed that a quantitative science of society might be quite amenable with conservative goals.

De Decker's words and Quetelet's defence are also reminders that *physique sociale* and *l'homme moyen* were never intended as epistemological tools, but terms Quetelet adopted to implement a plan of study. Rather than a theory meant to be tested against Durkheim's autonomous society or a failed path on the road to evolution, the only coherent way to frame Quetelet's work is as a methodological plan for individuals to serve government and scientific institutions. Individuals determined *physique sociale*, not the other way around. As de Decker foresaw, the ramifications for the 'average person' were felt not in the tyranny of numbers or a statistical determinism but through the self-sacrifice of will demanded of the practitioners of *physique sociale* – the real average men. Though not nearly as famous as these other thinkers, de Decker knew Quetelet well and shared many of the same assumptions about the necessity of stability in Belgium. His words also echoed the Catholicism and scepticism of human reason found in Quetelet's romantic heroes like Chateaubriand. The abstract threat to individuality posed by Dostoevsky's frightening 'percentages', seen in this reading, then becomes an illusion. The true threat was to anyone who wanted to participate in the process of *physique sociale* and therefore needed to abandon that part which was not suited to running a state or collecting information about human behaviour. The force of the numbers, it seemed, was felt most strongly by those who sought them.

CONCLUSION: THE NEW ARGONAUTS

With the perspective of over 150 years, much of the debate over free will and social physics seems moot. Even had Quetelet *intended* his science to be a predictive tool for individual action, and therefore wanted to efface all free will, by the late stages of his career the natural and human sciences were offering up theories with *far* greater ramifications for the autonomy of individual action than *physique sociale*. What great threat to human will could *l'homme moyen* offer in the wake of reflex studies, neurophysiology, psychology or evolutionary biology?[1] Contrasted with the great deal of attention devoted to the determinism question by James Clerk Maxwell, who while setting out the laws of physics also searched in physical laws for the possibility of 'free will' in stable systems, or the assertions of radical empiricists during the 1870s in England, Quetelet's occasional rhetorical drifts towards a deterministic social science seem almost quaint. In a rich irony, Maxwell and others would eventually even locate in Quetelet's statistics the possibility of *indeterminism*, as Catholics in particular saw a possibility in thermodynamics and Brownian motion to reconcile science and religion.[2] Had Quetelet not been swept up into the general outrage over Buckle's book, perhaps others would have come to the same conclusions about social physics as de Decker.

Removing Quetelet's social physics from the debates over free will or the grand intellectual tradition of Laplace and probability theory does not impoverish its legacy; rather, it returns Quetelet's ideas as close as possible to the way in which he understood them and the contexts in which they were developed. Neither does the perspective offered in this book lessen the importance for a historical understanding of the development of the first quantitative science of man; instead, it allows some of the more practical entailments of Quetelet's ideas to come into focus. It is ultimately a question left to each individual whether one is less free because people and institutions count individual acts, make nuerological scans of the brain, design logarithms to know human desires or trace the environmental adaptations of our prehistoric ancestors; as Montesquieu showed in *The Persian Letters*, freedom is, to a large degree, a matter of perspective. It *is* possible, however, that by recognizing Quetelet's social physics and *l'homme moyen* as potential models for how to serve bureaucratic institutions, rather than as fully realized philosophical ideas, some insight may gained about the lives of modern men and women who today serve institutions of state and science. If few critics of the nineteenth century recognized the consequences of social physics for individuals who chose to work in institutions where quantitative data on human behaviour was gathered, stored and analysed for information or profit, it may have been because there were so few of these workers at the time. Today that excuse no longer holds.

Investigating the practical consequences of social physics also does not leave Quetelet out of the larger story of the nineteenth-century social sciences. His ideas still were an 'intellectual odyssey' of the nineteenth century, just not always in the ways we might expect. Though the notion that he was merely an implementer of a Laplacian deterministic scheme is incorrect, he did follow in the tradition of the French sciences in many other ways. Consider, for example, the following two quotes from Laplace:

> Wise governments, convinced of the utility of learned societies, and viewing them as one of the principal foundations of the glory and of the prosperity of empires, have instituted and placed them near themselves to enlighten by their wisdom, from which they have frequently derived great advantage.[3]

> Let us apply to the political and moral sciences the method of observation upon calculus, the method which has served us so well in the natural sciences.[4]

Quetelet's biographers and subsequent historians have often pointed to the second of these two quotations to explain Laplace's influence on Quetelet's *physique sociale*. And it is true that Quetelet featured this quotation as the frontispiece to *Sur l'homme* in recognition of the importance of collecting data on human activity; if one were forced to make such a connection, the ideas of Laplace seem to lead most naturally to the ideas of Quetelet. Yet the first quotation, as hopefully demonstrated in this book, may better explain the importance of Laplace to Quetelet's career. Though a talented enough mathematician, the young Belgian certainly came closer to imitating Laplace in convincing 'governments' of the 'utility of learned societies' than in his scientific discoveries. There is a reason his statue stands today outside of the Académie royale, overlooking the Parc Bruxelles, steps away from the Palais royale and dozens of government buildings (see Figure 1.1). Social physics was never far from institutions of the state.

Even at the level of social thought, the Laplacian programme had few similarities in theory or practice to what Quetelet later developed. For Laplace, it was the theory – the theory of probabilities – which mattered, with the empirical data only necessary to confirm the ideas of the theory. But for Quetelet, it was almost certainly the opposite: the plan to gather data was paramount, and the theory only an afterthought. To invent a quarrel, it would seem the two men disagreed over what exactly was meant by the phrase 'methods of observation and calculation' in the second quotation. These 'methods' for Laplace had meant the sophisticated mathematical techniques which had proved so successful in astronomy, but for Quetelet the primary 'methods' of science flowed from data collectors directed by large and organized bodies of researchers. The instruction and direction of these researchers – both the leaders and workers of science institutions – which he had observed in Paris and Germany, and had instituted in Belgium, were the primary methodologies through which Quetelet built social physics.

The importance of institutions highlights the fact that the average man of science was not a creation of Quetelet alone but also of the time and place in which he lived. Quetelet's experience in Ghent as a poet and mathematician had not prepared him for the new organizational hierarchies of science, and it was through his personality more than specific scientific expertise that he managed to find success in appealing to King William and his advisors. The gentleman researchers at the Académie royale and Quetelet's frustrated friends were good reminders that not all men of science had been able to navigate this path and that those who tried to practice science as a form of *belles-lettres* or natural history would have little role in the sciences to come. Belgium may have held onto the Enlightenment consensus longer than other nations, but the reforms of the United Kingdom of the Netherlands meant that the 'war of the arts and sciences' came to the Low Countries as well. As Quetelet became exposed to the workings of a modernizing state, and watched his friends struggle to survive in the aftermath of the war, he may have realized he had better prospects in following administrative geniuses like Falck rather than the generations of quantitative thinkers that had inspired him like Condorcet, d'Alembert and Pascal, or the Romantic radicals like his heroes Chateaubriand and Madame de Staël. Interestingly, Quetelet ended up both between and outside of these traditions. Though the break in the Enlightenment consensus could have been a disaster for the broadly educated young Belgian, it ended up creating an entirely new path of research for which Quetelet was perfectly suited.

Outside of Belgium, Quetelet found confirmation in the reports from his correspondents and travels to France and Germany, the two nations that had helped to shape and reshape his country so many times. In the administrative practices of the Paris observatory, he learned not only the new ideas of probability emerging from the mathematicians like Poisson, Fourrier and Laplace, but also the practical demands and requirements for directing a large observatory from mentors like Bouvard and Arago. In Prussia, he learned from directors like Olbers that significant state support was crucial to directing a large observatory, and that the German practice of *statistik* had as much to contribute to the mathematical side of statistics as what he had learned in France. Even here it seems, in the earliest formations of *physique sociale* in the 1820s, Quetelet managed to find an average between the French and German nations that had made Belgium 'a nation of frontiers'.

The broad network of administrators, men of science and observers Quetelet encountered in the 1820s in Ghent, Brussels, Paris and Germany undoubtedly deserve to share credit with Quetelet in the creation of *physique sociale*. The first formulation of a true statistical science of society had required a singular figure like Quetelet who possessed the rare combination of a well-rounded background in the ideas of Enlightenment, technical training in mathematics and a personality inclined both to Romantic ideals and bureaucratic wrangling. But Quetelet would have been the last person to say that he created the science by himself. A true social physics was to be built with the men of state and science he found in the early administration of the United Kingdom of the Netherlands and in the pages of the *Correspondance mathématique et physique* and not by singular giants of scientific genius. Though largely ignored by most critics of *physique sociale*, it was these new men of science who mattered most to the future of quantitative studies of human behaviour.

Such was the plan of *Sciences mathématique et physique chez les belges*, the book quoted at the outset of this study, where Quetelet lauded the 'innovation of great importance' in the sciences that would allow Belgium to become a leader in European science. Quetelet had argued that Belgium lagged behind the great powers and was unable to compete in an era of great geniuses, yet Quetelet's experiences in Belgium and throughout the world of European science in the 1820 and '30s confirmed his belief that science could no longer progress through the individual genius or the lone researcher. No more clearly could this have been seen than in Germany where the model observatories and conferences were gradually pushing out the kind of science practised by his idol Goethe. In this new world of scientific work, Laplace's École polytechnique was a better model for scientific production than the sociable *gens de lettres* Quetelet first found at the Académie royale, and the kind of contemplative science practised in studies and salons moved to the regimented world of the observatory. Quetelet's friends in Ghent struggled too as the new men of the academy rose up through the ranks as diligent observers. As Frank Manuel wrote of the Saint-Simonians in France, 'the eighteenth century *philosophes* had respected the sciences and arts equally' but 'labor' had been raised 'to a new height ... [with] the commandment: "All men must work"'.[5] Though only a few heeded the call to join Saint-Simon's cult, many more signed up to work in the scientific institutions Quetelet was creating a few decades later.

By the second half of the nineteenth century, the workers had begun to show up in Belgium. Quetelet's belief, quoted in the introduction, that the 'man of talent ... ceased to act as an individual' in the nineteenth-century sciences, was one of his most important insights. Not only did this language echo de Decker's understanding of statecraft and science, where all those willing to serve state and science must abandon their personality and even their free will. It also paved the way for a nation like Belgium to take a role in the sciences of the future. Anticipating the later centrality of Brussels to European bureaucracy by over a century, Quetelet successfully brought the first International Maritime Conference to Brussels in 1853, the same year in which he hosted the first General

Statistical Congress, uniting both the quantitative and qualitative aspects of statistics. More projects may have followed had he not suffered a stroke in 1855, but the idea of Belgium taking its part in international science had been Quetelet's dream since the observatory project, a dream he pursued until his death in 1874 at age seventy-seven. The 'single observer', he claimed, needed to be replaced by 'active observers spread out across the globe', observers whom Quetelet believed would be called the 'New Argonauts ... [had the] ... era been more poetic'.[6] While as a small country Belgium may have been short on individual genius, the country would be able to contribute many New Argonauts to the cause of science.

Quetelet's history of the development of Belgian and European science as a progression from solitary genius to institutional networks of dedicated observers might help to explain why a young Belgian trained initially in geometry and the literary arts would take to such fields as statistics, meteorology, terrestrial magnetism, demography and *physique sociale*. Quetelet, seeing little chance of advancement in the sciences or arts as traditionally understood, seemed rarely to seek out opportunities that required solitary work or presented the chance to test potential scientific theories through creative experiment. As Quetelet learned in his pursuit of an observatory and confirmed in his meeting with observatory directors, institutions should precede ideas. The view of Quetelet dominant since his early twentieth-century biographies – that he had structured his work around the pursuit of one idea or another in an imperfect form – appears then to be somewhat misjudged; rather than following Laplace or Condorcet, or anticipating Durkheim or Galton, Quetelet followed the path of whichever discipline offered the best opportunity to engage in large-scale data collection and international collaboration. In other words, he was drawn to whichever disciplines were most disposed to take advantage of those '*savants* who had ceased to be individuals', the other average men of *physique sociale* who would redeem Belgium.

The project to elevate Belgium to the ranks of the elite scientific nations must be considered a success, as seen by the increasing international prominence of the Brussels Observatory after its completion in 1835. While Quetelet occasionally returned to population statistics and *physique sociale*, after this point his original research was directed towards fields that allowed for even greater large-scale collaboration than population statistics. After the observatory was completed, he published only three papers on statistics in comparison to over fifty on the earth sciences and astronomy. A large majority of these papers concerned the science that so delighted he and Gauss in Göttingen, terrestrial magnetism, which by the 1830s virtually mandated global observations, quantification and international collaboration.[7] Magnetism research had got its start at the Paris Observatory, and one of its greatest nineteenth-century advocates had been Quetelet's friend and mentor there during his 1823 travels, François Arago. Arago in turn had been part of an enterprise which 'typified the enthusiasm for data-collecting'.

What Edward Sabine (another of Quetelet's associates) had called the 'Magnetic Crusade' included not just the Frenchman Arago, but globe-hopping men of science Quetelet knew like Humboldt.[8] Yet as John Cawood has pointed out, magnetism moved in directions less suitable to 'Humboldtian science', with quantification pioneered by Gauss edging out the 'naturalistic approach to science'.[9] Though Quetelet had met and admired Humboldt for his leadership, he had little trouble adapting the observatory for use in this new global enterprise.

The project to count natural, rather than social, phenomena succeeded in other ways, though not always in the manner in which Quetelet predicted. John Herschel, another enthusiast for magnetism and close friend, helped Quetelet establish the Brussels Observatory by offering to collaborate on meteorological observations from the Cape of Good Hope.[10] It has been argued that Quetelet's old assistant Charles Morren, who had helped collect the Brussels population data, became the 'Father of Phenology', the science of the 'life cycles' of plants that has been crucial to much recent work on climate change.[11] The 2007 edition of the Intergovernmental Panel on Climate Change (IPCC IV) has also acknowledged that the 'international coordination of meteorological observations' began at the 1853 Maritime conference. In 1861 Quetelet collected nearly all the meteorological data and what he called 'periodic phenomena' into a book called *Sur le physique de globe*, a work as all encompassing for the life of the planet as *Physique sociale* had been for social life. It is a work that still awaits its incorporation into the historiography of climate science. In all, Quetelet managed to establish, in the eyes of at least one historian, Brussels as the 'one of the capitals of the earth sciences' in the years after the observatory was completed.[12] Based on the success of these projects and others, Quetelet may have proved the point he was so hard pressed to make in 1824 to the sceptical minister of education Van Ewyck.[13] Not only had discoveries been made, but his observatory had helped to start the nascent science of climatology.

The observatory was just one way in which Quetelet's New Argonauts could solidify Belgium's standing in European affairs; the myriad international conferences, congresses and associations that met in Brussels were another path towards internationalism.[14] Quetelet had known the disruptions caused by revolution and by war: he had grown up in French-occupied Belgium, had several friends who had fought in the Napoleonic Wars and the 1830 revolution had caused him personal and professional pain, keeping him away from his wife and delaying the observatory construction. It may be too much to attribute Quetelet's internationalism and search for statistical regularity to trauma in the Low Countries, but he was neither the first nor the last Belgian to seek for his country a central, mediating, role among the great powers of Europe. From Charles V to the visionary twentieth-century data collector Paul Otlet, Belgians have been at the forefront of internationalism.[15] Quetelet was, however, one of the first to propose that this mediating role in European affairs be played by men dedicated to serving institutions, concerned more with process and maintaining stability than with the 'personal and humane' elements described by de Decker. Tragically, Belgium did not succeed in preventing further disruption to its country in the twentieth century, but in creating a statistical science of man and a model for scientific work to supplant the era of the 'man of talent', both of Quetelet's average men have thrived in the nearly 150 years since his death. *L'homme moyen* as a means to understand human behaviour may eventually join some of the more famous theories of the nineteenth century as a curiosity in social thought, but as long as the average man of science endures – and it seems he is cultivated as assiduously in our time as he was in the nineteenth century – Adolphe Quetelet's social physics will remain.

EPILOGUE: THE AVERAGE ENLIGHTENMENT

Though the majority of this book has been devoted to a single Belgian's role in science over a period of a few decades, the questions that first led to a study of Quetelet are somewhat larger. They concern nothing less than the legacy of the Enlightenment, one of the most disputed periods of study for European history in general and the history of science in particular. Nothing that can be said about Quetelet can settle any of these debates, but attention to Quetelet and figures like him might introduce a few new questions: did the writers of the French Enlightenment intend to inspire mediocrity when they called for egalitarianism? Were the sciences of man, founded on an understanding of human affairs modelled on the study of the natural sciences, meant to be staffed with Quetelet's average men? Did the 'odyssey' of the nineteenth-century quantitative social sciences, the journey away from the progressivism of Condorcet towards Galton's scientific racism, wind its way through the offices of statistical societies, international congresses and other large-scale collaborative research institutions?

Such questions have largely been outside of the most heated debates in the 'Enlightenment wars', which have more often than not been fought at the extremes of political rhetoric. In the words of one avowed combatant, the contest has been over the fundamental question: 'Has the Enlightenment betrayed us or does it remain out best hope?'[1] While most popular discourse and history textbooks largely and reflexively praise the thinkers of the eighteenth century, there is a long strain of critical commentary that views the Enlightenment as the source of imperialism, totalitarianism, Western cultural superiority and other general calamities of recent history.[2] Some historians have even taken a very different tack, arguing that maybe the Enlightenment was not so important after all.[3] Critics, it seems, have either believed that the eighteenth-century project to study society imposed a historically unthinkable level of oppression on its objects of study ... or that it was a mostly innocuous project confined to a few elites.

Quetelet's professional career and social theory may suggest an alternative narrative of the Enlightenment legacy, one that unfortunately may not be as provocative (or triumphant) as earlier accounts, but one that has the benefit of retaining the idea of Enlightenment as a justifiable object of historical inquiry.

In this vision, the project to construct political and social knowledge on the sure footing of the natural sciences was successful, though in ways the Enlightenment authors could never have imagined; in placing their hopes in the tools of the natural sciences, these writers could not have realized they were constructing the end of the consensus between the arts and science. The *philosophes*, it might be said, made a world that valued science, but the science was not of their choosing; future generations of science have continued to weigh upon them. Who after all could have known in the eighteenth century what scientific research would have looked like 200 years later? Was it obvious in calling for a view of the world based on Newtonian physics that physics itself would be upended at the beginning of the twentieth century? And, most relevant to Quetelet's story, could it have been predicted that the project to create a science of man would lead to a proliferation of peer-reviewed studies created by professionals trained in the proper way to count? Not all of the social sciences or humanities have adopted a quantitative methodology, but one would be hard pressed to find a field that takes human activity as its object of study that does not have its numerate champions. For better or worse, the Enlightenment is alive in every moving, meaningful and absurd study that seeks quantitative correlations between human environment and behaviour.[4]

While Quetelet's average man of science may not have the dramatic appeal of Bentham's Panopticon or other dystopic bogeymen, it would seem to be far more emblematic of the extension of Enlightenment thought into understanding social behaviour.[5] Recognition of social theory as actually practised by those influenced by the Enlightenment, rather than tendentious associations between superficially similar forms of thought, would seem to be the better historical project. It may seem remarkable given the hold the Enlightenment has over modern thought that there have been relatively few historical studies of how Enlightenment ideas have been put into practice by nineteenth-century men of science, but such has been the consequence of narrow disciplinary work and politically motivated polemics.[6] Especially given the importance of science in the formulation of the Enlightenment project, it would seem a good place to start in understanding the legacy of the Enlightenment is to study the men of science who adopted the project of the *philosophes*, no matter how common their actions and activities may seem.[7]

WORKS CITED

Alembert, J. la R. d', 'Essai sur la société des gens de lettres et des grands', in J. la R. d'Alembert, *Oeuvres Complètes de d'Alembert* (1753; Geneva: Slatkine, 1967), pp. 335–72.

André, R., 'Adolphe Quetelet, académicien', *Actualité et universalité de la pensée scientifique d'Adolphe Quetelet: Actes du colloque organisé à l'occasion du bicentenaire de sa naissance* (Brussels: Académie royale de Belgique, 1997), pp. 23–45.

[Anon.], 'Journal des séances', *Nouveaux mémoire de l'Académie royale des sciences et des belles-lettres de Bruxelles*, 3 (1826), pp. i–xxiv.

[Anon.], 'Frauhofer – Refractive and Dispersive Powers of Glass, and the Achromatic Telescope', *Foreign Quarterly Review*, 1 (1827), pp. 424–35.

[Anon.], 'Memoir of the Life of Joseph Fraunhofer', *Journal of the Franklin Institute*, 4 (1829), pp. 96–103.

[Anon.], 'Review of *Recherches sur le penchant au crime aux différens âges*', *Bulletin des sciences géographiques, économique politique, voyages*, 28 (1831), pp. 113–36.

[Anon.], *Examination of Buckle's History of Civilization in England by A Country Clergyman* (London: Thomas Hatchard, 1858).

[Anon.], 'History of Civilization in England', *Crayon*, 4 (1857), pp. 336–8.

Arendt, H., *Eichmann in Jerusalem* (New York: Viking Press, 1963).

—, *The Origins of Totalitarianism* (New York: Harcourt Brace, 1973).

Ariew, A., 'Under the Influence of Malthus's Law of Population Growth: Darwin Eschews the Statistical Techniques of Adolphe Quetelet', *Study of History and Philosophy of Biological and Biomedical Sciences*, 38 (2007), pp. 1–19.

Ashworth, W. J., 'The Calculating Eye: Baily, Herschel, Babbage and the Business of Astronomy', *British Journal for the History of Science*, 27 (1994), pp. 409–41.

Aubin, D., 'The Fading Star of the Paris Observatory in the Nineteenth Century: Astronomers' Urban Culture of Circulation and Observation', *Osiris*, 18 (2003), pp. 79–100.

Baker, K. M., *Condorcet: From Natural Philosophy to Social Mathematics* (Chicago, IL: University of Chicago Press, 1975).

—, *Inventing the French Revolution: Essays on French Political Culture in the Eighteenth Century* (Cambridge: Cambridge University Press, 1990).

Bakich, M., *The Cambridge Planetary Handbook* (Cambridge: Cambridge University Press, 2000).

Barkan, E., *The Retreat of Scientific Racism* (Cambridge: Cambridge University Press, 1993).

Baron, A., 'Preface', in A. Baron (ed.), *Collection d'opuscules philosophique et littéraires* (Brussels: Wahlen, 1840), pp. iii–x.

Becker, B. J., *Unravelling Starlight: William and Margaret Huggins and the Rise of the New Astronomy* (Cambridge: Cambridge University Press, 2011).

Becker, P., and W. Clark, 'Introduction', in P. Becker and W. Clark (eds), *Little Tools of Knowledge: Historical Essays on Academic and Bureaucratic Practices* (Ann Arbor, MI: University of Michigan Press, 2001).

Bee, P., 'The BMI Myth', *The Guardian* (27 November 2006), p. 18.

Beirne, P., 'Adolphe Quetelet and the Origins of Positivist Criminology', *American Journal of Sociology*, 92 (1987), pp. 1140–69.

Bennett, J. A., *The Divided Circle: A History of Instruments for Astronomy, Navigation and Surveying* (Oxford: Phaidon-Christie's, 1987).

—, 'The English Quadrant in Europe: Instruments and the Growth of Consensus in Practical Astronomy', *Journal for the History of Astronomy*, 23 (1992), pp. 1–14.

Bergmans, P., 'Quetelet poète', *Ciel et terre*, 45 (1929), pp. 12–16.

Biagioli, M., *Galileo, Courtier: The Practice of Science in the Culture of Absolutism* (Chicago IL: University of Chicago Press, 1993).

Bigg, C., 'Staging the Heavens: Astrophysics and Popular Astronomy in the Late Nineteenth Century', in C. Bigg, D. Aubin and H. O. Sibum (eds), *The Heavens on Earth: Observatories and Astronomy in Nineteenth-Century Science and Culture* (Durham, NC: Duke University Press, 2010), pp. 305–24.

Bigg, C., D. Aubin and H. O. Sibum (eds), *The Heavens on Earth: Observatories and Astronomy in Nineteenth-Century Science and Culture* (Durham, NC: Duke University Press, 2010).

Bigg, C., D. Aubin and H. O. Sibum, 'Introduction: Observatory Techniques in Nineteenth-Century Science and Society', in C. Bigg, D. Aubin and H. O. Sibum (eds), *The Heavens on Earth: Observatories and Astronomy in Nineteenth-Century Science and Culture* (Durham, NC: Duke University Press, 2010), pp. 1–32.

Bolus-Reichert, C., *The Age of Eclecticism* (Columbus, OH: Ohio State University Press, 2009).

Bonald, L. de, 'Des sciences, des lettres et des arts', in L. de Bonald, *Oeuvres de M. Bonald: Mélanges littéraires, politiques et philosophiques*, 8 vols (1807; Brussels: Gérant, 1845), vol. 7, pp. 349–80.

—, 'Sur la guerre des sciences et lettres', in L. de Bonald, *Oeuvres de M. Bonald: Mélanges littéraires, politiques et philosophiques*, 8 vols (Brussels: Gérant, 1845), vol. 7, pp. 459–64.

Bourdieu, P., 'The Peculiar History of Scientific Reason', *Sociological Forum*, 6 (1991), pp. 3–26.

—, *Les Règles de l'art: Genèse et structure du champ littéraire* (Paris: Seuil, 1992).

Bourgeois, J., 'Planet', J. L. Heilbron (ed.), *The Oxford Guide to the History of Physics and Astronomy* (Oxford: Oxford University Press, 2005), pp. 253–6.

Brian, É., 'Transactions statistiques au XIXe siècle', *Actes de la recherche en sciences sociales*, 145 (2002), pp. 34–46.

Brooks, D., *The Social Animal: The Hidden Sources of Love, Character, and Achievement* (New York: Random House, 2011).

Browne, J., *Charles Darwin: The Power of Place* (New York: Alfred A. Knopf, 2003).

Brush, S. G., *The History of Modern Science: A Guide to the Second Scientific Revolution* (Ames, IA: Iowa State University Press, 1988).

Buchnan, J., *Crowded with Genius: The Scottish Enlightenment, Edinburgh's Moment of the Mind* (New York: Harper Collins, 2003).

Buck, P., 'From Celestial Mechanics to Social Physics: Discontinuity in the Development of the Sciences in the Early Nineteenth Century', in H. N. Janke and M. Otte (eds), *Epistemological and Social Problems of the Sciences in the Early Nineteenth Century* (Dordrecht: D. Reidel, 1981), pp. 19–33.

Buckle, H. T., *History of Civilization in England*, new edn, 3 vols (1857; London: Longmans, Green, and Co., 1885).

Bulmer, M., *Francis Galton: Pioneer of Heredity and Biometry* (Baltimore, MD: Johns Hopkins University Press, 2003).

Cannon, S. F., *Science in Culture: The Early Victorian Period* (New York: Science History Publications, 1978).

Caplan, J., and J. Torpey (eds), *Documenting Individual Identity: The Development of State Practices in the Modern World* (Princeton, NJ: Princeton University Press, 2001).

Carter, C., *Magnetic Fever: Imperialism and Empiricism in the Nineteenth Century* (Philadelphia, PA: American Philosophical Society, 2009).

Cassirer, E., *The Philosophy of the Enlightenment*, trans. F. C. A. Koelln and J. P. Pettegrove (German, 1932; Princeton, NJ: Princeton University Press, 1951).

—, *Determinism and Indeterminism in Modern Physics: Historical and Systematic Studies of the Problem* (New Haven, CT: Yale University Press, 1956).

Cassis, Y., *Capitals of Capital: The Rise and Fall of International Financial Centres, 1780–2009*, new edn (2006; Cambridge: Cambridge University Press, 2010).

Cawood, J., 'Terrestrial Magnetism and the Development of International Collaboration in the Early Nineteenth Century', *Annals of Science*, 24 (1977), pp. 551–87.

—, 'The Magnetic Crusade: Science and Politics in Early Victorian Britain', *Isis*, 70 (1979), pp. 492–518.

Chateaubriand, F.-R. de, 'Essai sur les révolutions', in F.-R. de Chateaubriand, *Essai sur les révolutions, Génie du christianisme*, ed. M. Regard (1797; Paris: Gallimard, 1978).

—, 'Lettre au C. Fontanes sur la seconde édition de l'ouvrage de Madame de Staël', *Mercure de France*, 3 (1801), pp. 14–38.

—, 'Génie du christianisme', in F.-R. de Chateaubriand, *Essai sur les révolutions, Génie du christianisme*, ed. M. Regard (1802; Paris: Gallimard, 1978).

—, *Mémoires d'outre-Tombe*, 4 vols (1849–50; Paris: Garnier, 1998), vol. 2.

Cohen, I. B., *Revolution in Science* (Cambridge, MA: Belknap Press, 1985).

—, 'Revolutions, Scientific, and Probabilistic', in L. Krüger, L. J. Daston and M. Heidelberger (eds), *The Probabilistic Revolution*, 2 vols, vol. 1: *Ideas in History* (Cambridge, MA: Massachusetts Institute of Technology Press, 1987), pp. 23–44.

Collard, A, 'Le Centenaire de la création de l'Observatoire royal de Bruxelles', *Ciel et terre*, 42 (1926), pp. 209–23.

—, 'Goethe et Quetelet: Leurs relations de 1829 à 1832', *Isis*, 20 (1934), pp. 426–35.

—, 'Adolphe Quetelet et l'astronomie', *Ciel et terre*, 6 (1935), pp. 1–10.

Collins, R., *The Sociology of Philosophies: A Global Theory of Intellectual Change* (Cambridge, MA: Belknap Press, 1998).

Comte, A., *Cours de philosophie positive*, 6 vols (Paris: Bachelier, 1839), vol. 4.

Condorcet, 'A General View of the Sciences of Social Mathematics', in *Condorcet: Selected Writings*, ed. and trans. K. M. Baker (French, 1793; Indianapolis, IN: Bobbs-Merrill, 1976), pp. 183–206.

Cooper, B. P., *Family Fictions and Family Facts: Harriet Martineau, Adolphe Quetelet, and the Population Question in England* (New York: Routledge, 2007).

Cooper, B. P. and M. S. Murphy, 'The Death of the Author at the Birth of Social Science: The Cases of Harriet Martineau and Adolphe Quetelet', *Studies in History and Philosophy of Science Part A*, 31 (2000), pp. 1–36.

Cousin, V., *Cours de philosophie, professé à la faculté des lettres pendant l'année 1818 ...* (Paris: Hachette, 1836).

—, *Fragments de philosophie cartésienne* (Paris: Didier, 1852).

Crosland, M., *Science under Control: The French Academy of Sciences, 1795–1914* (1992; Cambridge: Cambridge University Press, 2002).

—, 'A Science Empire in Napoleonic France', *History of Science*, 44 (2006), pp. 29–48.

Darnton, R., *The Literary Underground of the Old Regime* (Cambridge, MA: Harvard University Press, 1982).

Daston, L., *Classical Probability in the Enlightenment* (Princeton, NJ: Princeton University Press, 1988).

—, 'Science Studies and the History of Science', *Critical Inquiry*, 35 (2009), pp. 798–813.

Daston, L., and P. Galison, *Objectivity* (New York: Zone, 2007).

Daumas, M., *Scientific Instruments of the Seventeenth and Eighteenth Century*, ed. and trans. M. Hobrook (French, 1972; New York: Praeger, 1972).

—, *Arago, 1786–1853: La Jeunesse de la science*, 2nd edn (1987; Paris: Belin, 1987).

Dear, P., and S. Jasanoff, 'Dismantling Boundaries in Science and Technology Studies', *Isis*, 101 (2010), pp. 759–74.

Decker, P. de, 'De l'influence du libre arbitre de l'homme sur les faits sociaux', *Mémoires de l'Académie royale des sciences, des lettres et des beaux-arts de Belgique*, 21 (1848), pp. 69–92.

Demoulin, M. A., 'Adolphe Quetelet, fondateur de l'Observatoire royal de Belgique', *Bulletin astronomique de l'Observatoire royal de Belgique*, 2 (1935), pp. 1–3.

Derrida, J., *The Archaeology of the Frivolous: Reading Condillac*, trans. J. P. Leavey, Jr, (French, 1973; Pittsburgh, PA: Duquesne University Press, 1980).

Desrosières, A., 'Quetelet et la sociologie quantitative', *Actualité et universalité de la pensée scientifique d'Adolphe Quetelet: Actes du colloque organisé à l'occasion du bicentenaire de sa naissance* (Brussels: Académie royale de Belgique, 1997), pp. 179–98.

—, *The Politics of Large Numbers: A History of Statistical Reasoning*, trans. C. Nash (French, 1993; Cambridge, MA: Harvard University Press, 1998).

Dewalque, G., 'François-Philippe Cauchy', in *Biographie nationale* (Brussels: Bruylant, 1872), vol. 3, pp. 380–3.

Dhombres, J., 'Preface', in M. Daumas, *Arago, 1786–1853: La Jeunesse de la science*, 2nd edn (Paris: Belin, 1987), pp. 5–9.

Dickens, C., *Hard Times. For these Times* (London: Bradbury & Evans, 1854).

Donder, L. W.-de, 'Une Lettre de Cécile Quetelet relative à la révolution Belge de 1830', *Cahiers bruxellois*, 6 (1961), pp. 68–75.

—, *Inventaire de la correspondance d'Adolphe Quetelet déposé à l'Académie royale de Belgique* (Brussels: L'Académie royale de Belgique, 1968).

Donnelly, K., 'Social Physics or Social Disease? Quetelet, Villermé, and Cholera in Brussels and Paris, 1832', in E. Martone (ed.), *Royalists, Radicals, and les Misérables: France in 1832* (Newcastle-upon-Tyne: Cambridge Scholars Publishing, 2013), pp. 99–119.

—, 'On the Boredom of Science: Positional Astronomy in the Nineteenth Century', *British Journal for the History of Science*, 47 (2014), pp. 473–503.

—, 'The Other Average Man: Science Workers in Quetelet's Belgium', *History of Science*, 52 (2014), pp. 401-28.

Dostoevsky, F., *Crime and Punishment*, trans. R. Pevear and L. Volokhonksy (Russian, 1867; London: Vintage, 2007).

Dunnington, G. W., *Carl Friedrich Gauss: Titan of Science* (New York: Mathematical Association of America, 1955).

Durkheim, E., *Suicide: A Study in Sociology*, ed. and trans. J. A. Spaulding and G. Simpson (French, 1897; Glencoe, IL: Free Press, 1951).

Eknoyan, G., 'Adolphe Quetelet (1796–1874) – the Average Man and Indices of Obesity', *Nephrol Dial Transplant*, 23 (2008), pp. 47–51.

Espy, J. P., *To the Friends of Science* (n.p., 1845).

Fara, P., *Newton: The Making of Genius* (New York: Columbia University Press, 2007).

Fidanza, F., et al., 'Indices of Relative Weight and Obesity', *Journal of Chronic Diseases*, 25 (1972), pp. 329–43.

Foucault, M., 'What Is Enlightenment?', in P. Rabinow (ed.), *The Foucault Reader* (New York: Pantheon, 1984), pp. 32–50.

Fox, R., *The Savant and the State: Science and Cultural Politics in Nineteenth-Century France* (Baltimore, MD: Johns Hopkins University Press, 2011).

Francotte, H., 'Essai historique sur la propagande des encyclopédistes français à Liège', *Mémoires couronnés et autres mémoires*, 30 (1880).

Francken, E., and E. Mahaim, 'La Loi d'erreur de M. F. Y. Edgeworth', *Revue universelle des mines, de la métallurgie des travaux publics, des sciences et des arts appliqués à l'industrie*, 22 (1908), pp. 219–66.

Frickel., S., and N. Gross, 'A General Theory of Scientific/Intellectual Movements', *American Sociological Review*, 70 (2005), pp. 204–32.

Fuchs, S., 'A Sociological Theory of Scientific Change', *Social Forces*, 71 (1993), pp. 933–53.

Gieryn, T. F., 'Boundary-Work and the Demarcation of Science from Non-Science: Strains and Interests in Professional Ideologies of Scientists', *American Sociological Review*, 48 (1983), pp. 781–95.

Gigerenzer, G., et al., *The Empire of Chance: How Probability Changes Science and Everyday Life* (Cambridge: Cambridge University Press, 1989).

Gill, G., *We Two: Victoria and Albert: Rulers, Partners, Rivals* (New York: Random House, 2009).

Gillispie, C. C., 'Probability and Politics', *Proceedings of the American Philosophical Society*, 116 (1972), pp. 1–20.

—, *Science and Polity in France: The End of the Old Regime* (Princeton, NJ: Princeton University Press, 1980).

—, *Science and Polity in France: The Revolutionary Years* (Princeton, NJ: Princeton University Press, 2004).

Gillispie, C. C., with R. Fox and I. Grattan-Guinness, *Pierre-Simon Laplace 1749–1827: A Life in Exact Science* (Princeton, NJ: Princeton University Press, 1997).

Goldman, L., 'The Origins of British "Social Science": Political Economy, Natural Science and Statistics, 1830–1835', *Historical Journal*, 26 (1983), pp. 587–616.

Golinski, J., *Making Natural Knowledge: Constructivism and the History of Science* (Cambridge: Cambridge University Press, 1998).

Golinski, J., W. Clark and S. Schaffer (eds), *The Sciences in Enlightened Europe* (Chicago, IL: University of Chicago Press, 1999).

Gould, S. J., *The Mismeasure of Man (Revised and Expanded)* (1981; New York: W. W. Norton & Co., 2008).

I. Grattan-Guiness, *Convolutions in French Mathematics: From the Calculus and Mechanics to Mathematical Analysis and Mathematical Physics*, 3 vols (Basel: Birkhäuser, 1990).

Graunt, J., 'Natural and Political Observations upon the Bills of Mortality', in C. H. Hull (ed.), *The Economic Writings of Sir William Petty* (1665; Ithaca, NY: Cornell University Press, 1963), pp. 314–435.

Gray, J., *Enlightenment's Wake: Politics and Culture at the Close of the Modern Age* (London: Routledge, 1995).

Great Britain Geographical Section of the Naval Intelligence Division, *A Manual of Belgium and the Adjoining Territories* (London: His Majesty's Stationary Office, 1918).

Grégoire, H., 'Education in Belgium', in J.-A. Goris (ed.), *Belgium* (Berkeley, CA: University of California Press, 1945), pp. 226–38.

Hacking, I., 'Nineteenth Century Cracks in the Concept of Determinism', *Journal of the History of Ideas*, 44 (1983), pp. 455–75.

—, *The Emergence of Probability: A Philosophical Study of Early Ideas and Probability, Induction and Statistical Interference* (Cambridge: Cambridge University Press, 1984).

—, 'Prussian Numbers, 1860–1882', in L. Krüger, L. J. Daston and M. Heidelberger (eds), *The Probabilistic Revolution*, 2 vols, vol. 1: *Ideas in History* (Cambridge, MA: Massachusetts Institute of Technology Press, 1987), pp. 377–94.

—, *The Taming of Chance* (Cambridge: Cambridge University Press, 1990).

Hahn, R., *The Anatomy of a Scientific Institution: The Paris Academy of Sciences, 1666–1803* (Berkeley, CA: University of California Press, 1971).

Halbwachs, M., *La Théorie de l'homme moyen: Essais sur Quetelet et la statistique morale* (Paris: Librairie Félix Alcan, 1913).

Hall, T., *Carl Friedrich Gauss*, trans. A. Froderberg (Swedish, 1965; Cambridge, MA: Massachusetts Institute of Technology Press, 1970).

Hankins, F. H., *Adolphe Quetelet as Statistician* (New York: Columbia University Press, 1908), pp. 1–136.

—, 'Review of *Quetelet, statisticien et sociologue*', *Political Science Quarterly*, 27 (1912), pp. 719–22.

Hankins, T. L., *Jean d'Alembert: Science and the Enlightenment* (Oxford: Clarendon Press, 1970).

—, 'A Large and Graceful Sinuosity', *Isis*, 94 (2006), pp. 605–33.

Heilbron, J., *The Rise of Social Theory* (Cambridge: Polity Press, 1995).

Heilbron, J., N. Guilhot and L. Jeanpierre, 'Toward a Transnational History of the Social Sciences', *Journal of the History of the Behavioral Sciences*, 44 (2008), pp. 146–60.

Heilbron, J. L., T. Frängsmyr and R. E. Rider (eds), *The Quantifying Spirit in the 18th Century* (Berkeley: University of California Press, 1990).

Helvétius, C.-A., *De l'esprit* (Paris: Durand, 1758).

[Herschel, J.], 'Quetelet on Probabilities', *Edinburgh Review*, 92 (1850), pp. 1–57.

Heyck, H., 'The Organizational Revolution and the Human Sciences', *Isis*, 105 (2014), pp. 1–31.

Holton, G. J., P. Galison and S. S. Schweber (eds), *Einstein for the 21st Century: His Legacy in Science, Art, and Modern Culture* (Princeton, NJ: Princeton University Press, 2008).

Horkheimer, M., and T. W. Adorno, *Dialectic of Enlightenment: Philosophical Fragments*, trans. E. Jephcott (German, 1944; Stanford, CA: Stanford University Press, 2002).

Hulliung, M., *Citizen Machiavelli* (Princeton, NJ: Princeton University Press, 1983).

—, *The Autocritique of Enlightenment: Rousseau and the Philosophes* (Cambridge, MA: Harvard University Press, 1994).

—, 'Rousseau et les Philosophes: Facing up to the "Enlightenment Wars"', in M. O'Dea (ed.), *Rousseau et les Philosophes* (Oxford: Voltaire Foundation, University of Oxford Press, 2010).

Janiak, A., 'Newton and the Reality of Force', *Journal of the History of Philosophy*, 45 (2007), pp. 127–47.

Kern, S., *A Cultural History of Causality: Science, Murder Novels, and Systems of Thought* (Princeton, NJ: Princeton University Press, 2004).

Kevles, D. J., *In the Name of Eugenics: Genetics and the Uses of Human Heredity* (New York: Knopf, 1985).

Khosla, T., and C. R. Lowe, 'Indices of Obesity Derived from Body Weight and Height', *British Journal of Preventive and Social Medicine*, 21 (1967), pp. 122–8.

Klep, P. M. M., and I. H. Stamhuis, 'Introduction', in P. M. M. Klep and I. H. Stamhuis (eds), *The Statistical Mind in a Pre-Statistical Era: The Netherlands, 1750–1850* (Amsterdam: Askant, 2002).

Klinck, D., *The French Counterrevolutionary Theorist Louis de Bonald (1754–1840)* (New York: P. Lang, 1996).

Knapp, G. F., 'Die neuren Ansichten über Moralstatistik', *Jahrbücher für Nationalökonomie und Statistik*, 16 (1871), pp. 237–50.

Kohn, H., 'Introduction', in H. T. Buckle, *History of Civilization in England* (New York: Frederick Ungar, 1964).

Kossmann, E. H., *The Low Countries, 1780–1940* (Oxford: Clarendon Press, 1978).

Krüger, L., 'The Slow Rise of Probabilism', in L. Krüger, L. J. Daston and M. Heidelberger (eds), *The Probabilistic Revolution*, 2 vols, vol. 1: *Ideas in History* (Cambridge, MA: Massachusetts Institute of Technology Press, 1987), pp. 59–89.

Kuhn, T., 'The Function of Measurement in Modern Physical Sciences', *Isis*, 52 (1961), pp. 161–93.

—, *The Structure of Scientific Revolutions* (Chicago, IL: University of Chicago Press, 1962).

Kuntzinger, M. J., 'Essai historique sur la propagande des encyclopédistes français en Belgique dans la seconde moitié de XVIIIe siècle', *Mémoires couronnés et autres mémoires*, 30 (1880).

Lalvani, S., *Photography, Vision, and the Production of Modern Bodies* (Albany, NY: State University of New York Press, 1996).

Lamartine, A. de, *Des destinées de la poésie* (Paris: Gosselin, 1834).

Laplace, P.-S. de, *Exposition du système du monde*, 3rd edn (1796; Paris: Chez Courcier, 1808).

—, *Essai philosophique sur les probabilités* (1814; Brussels: Culture et Civilisation, 1967).

Larson, E. J., 'Public Science for a Global Empire: The British Quest for the South Magnetic Pole', *Isis*, 102 (2011), pp. 34–59.

Latour, B., *Science in Action: How to Follow Scientists and Engineers through Society* (Cambridge, MA: Harvard University Press, 1987).

Lalvani, S., *Photography, Vision, and the Production of Modern Bodies* (Albany, NY: State University of New York Press, 1996).

Lavalleye, J., *L'Académie royale des sciences, des lettres et des beaux-Arts, 1772–1972: Esquisse historique* (Brussels: Palais des Académies, 1972).

Lazerfeld, P. F., 'Notes on the History of Quantification in Sociology – Trends, Sources and Problems', *Isis*, 52 (1961), pp. 277–333.

Lécuyer, B.-P., 'Probability in Vital and Social Statistics: Quetelet, Farr, and the Bertillons', in L. Krüger, L. J. Daston and M. Heidelberger (eds), *The Probabilistic Revolution*, 2 vols, vol. 1: *Ideas in History* (Cambridge, MA: Massachusetts Institute of Technology Press, 1987), pp. 317–35.

Lepenies, W., *Between Literature and Science: The Rise of Sociology* (Cambridge: Cambridge University Press, 1988).

Levitt, T., '"I Thought This Might Be of Interest...": The Observatory of Public Enterprise', in C. Bigg, D. Aubin and H. O. Sibum (eds), *The Heavens on Earth: Observatories and*

Astronomy in Nineteenth-Century Science and Culture (Durham, NC: Duke University Press, 2010), pp. 285–304.

Lévy, P., 'Quetelet, poète et le statisticien', in P. Lévy, *Adolphe Quetelet, 1796–1874: Recueil des travaux et contributions présentés en 1974 en hommage à son rôle de statisticien* (Brussels: L'Académie royale de Belgique, 1974), pp. 43–56.

Locher, F., 'The Observatory, the Land-Based Ship and the Crusades: Earth Sciences in European Context, 1830–50', *British Journal for the History of Science*, 40 (2007), pp. 491–504.

Loise, F., 'Jean-François Lemaire', *Biographie nationale* (Brussels: Bruylant, 1891), vol. 11, pp. 779–81.

Lottin, J., 'Statistique morale et le déterminisme', *Revue néo-scolastique de philosophie*, 15 (1908), pp. 48–89.

—, 'Le Libre arbitre et les lois sociologiques d'après Quetelet', *Revue néo-scolastique de philosophie*, 18 (1911), pp. 479–515.

—, *Quetelet, statisticien et sociologue* (Louvain: Institut supérieur de philosophie, 1912).

Lucier, P., 'The Professional and the Scientist in Nineteenth-Century America', *Isis*, 100 (2009), pp. 699–732.

Luthereau, J. G. A., *Notice sur M. le Baron de Reiffenberg* (Brussels: Galarie St.-Hubert, 1850), p. 4.

MacLean, M., 'History in a Two-Cultures World: The Case of the German Historians', *Journal of the History of Ideas*, 49 (1998), pp. 473–94.

Mailly, E., 'Essai sur la vie et les ouvrages de Lambert-Adolphe-Jaques Quetelet', *Annuaire de l'Académie royale de Belgique* (1875), pp. 108–297.

Mackenzie, D., *Statistics in Britain, 1865–1930* (Edinburgh: Edinburgh University Press, 1981).

Malthus, T. R., *Malthus: 'An Essay on the Principle of Population'*, ed. D. Winch (1798; Cambridge: Cambridge University Press, 1992).

Manuel, F., 'From Equality to Organicism', *Journal of the History of Ideas*, 17 (1956), pp. 54–69.

—, *The New World of Henri Saint-Simon* (Cambridge, MA: Harvard University Press, 1956).

—, *The Prophets of Paris* (Cambridge, MA: Harvard University Press, 1962).

Mayr, E., *The Growth of Biological Thought: Diversity, Evolution, and Inheritance* (Cambridge, MA: Belknap Press, 1982).

Menand, L., *The Metaphysical Club* (New York: Farrar, Straus and Giroux, 2001).

Merz, J. T., *A History of European Thought in the Nineteenth Century*, 2 vols (Edinburgh: Blackwood, 1912).

Molvig, O., 'The Berlin Urania, Humboldtian Cosmology, and the Public', in C. Bigg, D. Aubin and H. O. Sibum (eds), *The Heavens on Earth: Observatories and Astronomy in Nineteenth-Century Science and Culture* (Durham, NC: Duke University Press, 2010), pp. 325–44.

Morehead, L. M., *A Few Incidents in the Life of Professor James P. Espy* (Cincinnati, OH: R. Clarke & Co., 1888).

[Morgan, C.], 'Review of Quetelet's *On Man*', *Athenaeum*, 406 (1835), pp. 593–5, 611–13, 658–61.

Morrell, J., and A. Thackray, *Gentleman of Science: Early Years of the British Association of the Advancement of Science* (Oxford: Oxford University Press, 1981).

Mosselmans, B., 'Adolphe Quetelet, the Average Man and the Development of Economic Methodology', *European Journal of the History of Economic Thought*, 12 (2005), pp. 565–82.

Nieuport, C., *Essai sur la théorie du raisonnent* (Brussels: Lemaire, 1805).

Norton, P. J., and B. J. Weiss, 'The Role of Courage on Behavioral Approach in a Fear-Eliciting Situation: A Proof-of-Concept Pilot Study', *Journal of Anxiety Disorders*, 23 (2009), pp. 212–17.

Nye, M. J., 'The Moral Freedom of Man and the Determinism of Nature: The Catholic Synthesis of Science and History in the *Revue des Questions Scientifiques*', *British Journal for the History of Science*, 9 (1976), pp. 274–92.

Oberschall, A., 'The Two Empirical Roots of Social Theory and the Probability Revolution', in G. Gigerenzer, L. Krüger and M. S. Morgan (eds), *The Probabilistic Revolution*, 2 vols, vol. 2: *Ideas in the Sciences* (Cambridge, MA: Massahusetts Institute of Technology Press, 1987), pp. 103–34.

Oliver, E. J., *Fat Politics: The Real Story Behind America's Obesity Epidemic* (Oxford: Oxford University Press, 2006).

Ore, O., *Cardano the Gambling Scholar* (Princeton, NJ: Princeton University Press, 1953).

Ormond, G. W. T., 'The Question of the Netherlands, 1829–1830', *Transactions of the Royal Historical Society*, 2 (1919), pp. 150–71.

Outram, D., 'Politics and Vocation: French Science, 1793–1830', *British Journal for the History of Science*, 12 (1980), pp. 27–43.

—, *Georges Cuvier: Vocation, Science, and Authority in Post-Revolutionary France* (Manchester: Manchester University Press, 1984).

—, *The Enlightenment* (Cambridge: Cambridge University Press, 1995).

Pachuri, R. K., and A. Reisenger (eds), *Climate Change 2007: Synthesis Report* (Geneva: International Panel on Climate Control, 2007).

Paepe, J.-L. de (ed.), *Exposition documentaire présentée à la Bibliothèque royale Albert I à l'occasion du centiare de la mort d'Adolphe Quetelet* (Brussels: Académie royale de Belgique, 1974).

Paller, B. T., 'Naturalized Philosophy of Science, History of Science, and the Internal/External Debate', *Proceedings of the Biennial Meeting of the Philosophy of Science Association*, 1 (1986), pp. 258–68.

Paperno, I., *Suicide as a Cultural Institution in Dostoevsky's Russia* (Ithaca, NY: Cornell University Press, 1997).

Pâquet, P., 'Les Initiatives de Adolphe Quetelet en astronomie', *Actualité et universalité de la pensée scientifique d'Adolphe Quetelet: Actes du colloque organisé à l'occasion du bicentenaire de sa naissance* (Brussels: Académie royale de Belgique, 1997), pp. 63–72.

Pascal, B., 'Discours sur les passions de l'amour', in B. Pascal, *Pascal: Oeuvres complètes*, ed. L. Lafuma (1652–3; Paris: Éditions du Seuil, 1963), pp. 285–9.

—, 'Préface sur le traité du vide' (1651), in B. Pascal, *Pascal: Oeuvres completes*, ed. L. Lafuma (1651; Paris: Éditions du Seuil, 1963), pp. 230–2.

Paul, H. W., *From Knowledge to Power: The Rise of the Science Empire in France, 1860–1939* (Cambridge: Cambridge University Press, 1995).

Pentland, A., *Social Physics: How Good Ideas Spread – Lessons from a New Science* (New York: Penguin Press, 2014).

Piot, C., 'Notice sur Pierre-Jacques-François De Decker', *Annuaire de l'Académie Royale*, 58 (1892), pp. 215–75.

Pirenne, H., *La Nation belge* (Brussels: Guyot, 1899).

Porter, T. M., 'The Mathematics of Society: Variation and Error in Quetelet's Statistics', *British Journal for the History of Science*, 18 (1985), pp. 51–69.

—, 'Lawless Society: Social Science and the Reinterpretation of Statistics in Germany, 1850–1880', in L. Krüger, L. J. Daston and M. Heidelberger (eds), *The Probabilistic Revolution*, 2 vols, vol. 1: *Ideas in History* (Cambridge, MA: Massachusetts Institute of Technology Press, 1987), pp. 351–75.

—, *The Rise of Statistical Thinking 1820–1900* (Princeton, NJ: Princeton University Press, 1986).

—, *Trust in Numbers* (Princeton, NJ: Princeton University Press, 1995).

—, 'Was Quetelet a Positivist?', *Actualité et universalité de la pensée scientifique d'Adolphe Quetelet: Actes du colloque organisé à l'occasion du bicentenaire de sa naissance* (Brussels: Académie royale de Belgique, 1997), pp 199–209.

Prévost, J.-G., and J.-P. Beaud, *Statistics, Public Debate and the State, 1800–1945: A Social, Political and Intellectual History of Numbers* (London: Pickering & Chatto, 2012).

Proctor, R., *Racial Hygiene: Medicine under the Nazis* (Cambridge: Cambridge University Press, 1988).

Quetelet, A, 'Essai sur la romance', *Annales Belgique*, 8 (1823), pp. 223–32, 295–311.

—, *Astronomie élémentaire* (1826), 3rd edn, 2 vols (Brussels: Tircher, 1834).

—, 'Mémoire sur les lois des naissances et de la mortalité à Bruxelles', *Nouveaux mémoires de l'Académie*, 3 (1826), pp. 493–512.

—, *Astronomie populaire* (1827), 2nd edn (Brussels: Remy, 1832).

—, *Recherches sur la population, les naissances, les décès, les dépôts de mendicité, etc. dans le Royaume des Pays-Bas* (Brussels: Tarlier, 1827).

—, *Instructions populaires sur le calcul des probabilités* (Brussels: Tarlier and Hayez, 1828).

—, *Recherches statistique sur le Royaume Des Pays-Bas* (Brussels: Hayez, 1829).

—, *Notes extraits d'un voyage scientifique fait en Allemagne pendant l'été de 1829* (Brussels: Hayez, 1830).

—, *Recherches sur la loi de la croissance de l'homme* (Brussels: Hayez, 1831).

—, *Recherches sur le penchant au crime aux différens âges*, 2nd edn (1831; Brussels: Hayez, 1833).

—, *Recherches sur le poids de l'homme aux différens âges* (Brussels: Hayez, 1833).

—, *Sur l'homme et le développement de ses facultés*, 2 vols (Brussels: Bachelier, 1835).

—, 'Preface', in *A Treatise on Man and the Development of his Faculties*, ed. T. Silbert, trans. R. Knox (French, 1835; Edinburgh: W. and R. Chambers, 1842), pp. v–x.

—, *Notice sur Alexis Bouvard* (Brussels: Hayez, 1844).

—, *Notice sur Antoine Reinhard Falck* (Brussels: Hayez, 1844).

—, *Lettres à S.A.R. le duc régnant de Saxe-Cobourg et Gotha: Sur la théorie des probabilités, appliquée aux sciences morales et politiques* (Brussels: Hayez, 1846).

—, *Notice sur le Colonel G-P Dandelin* (Brussels: Hayez, 1848).

—, *Du système social et des lois qui le régissent* (Paris: Guillaumin, 1848).

—, 'Sur la statistique morale et les principes qui doivent en former la base', *Nouveaux mémoires de l'Académie royale des sciences et belles-lettres de Bruxelles*, 31 (1848), pp. 3–68.

—, *Notice sur Louis-Vincent Raoul* (Brussels: Hayez, 1849).

—, *Notice sur Frédéric-August-Ferdinand-Thomas Baron de Reiffenberg* (Brussels: Hayez, 1852).

—, 'Quelques remarques sur l'influence des académies, des congrès, et des conférences scientifique' (1853), *Quetelet Varia (QV)*, Bibliothèque royale de Belgique, vol. 4, #9.

—, 'Dominique-François-Jean Arago', *Annuaire de l'Académie royale de Belgique* (1855).

—, *Sciences mathématique et physique chez les belges au XIXe siècle* (Brussels: Buggenhoudt, 1866).

—, *Physique sociale ou essai sur le développement des facultés de l'homme*, ed. É. Vilquin and J-P. Sanderson (1869; Brussels: Académie royale de Belgique, 1995).

—, *Des lois concernant le développement de l'homme* (Brussels: Hayez, 1870).

—, *Histoire des sciences mathématique et physique chez les belges*, new edn (1864; Brussels: Muquart, 1871).

Quetelet, A., and G. Dandelin, 'Jean-Second Opera: En une acte et en prose', Quetelet Archives, Académie royale de Belgique, Brussels (ARB), file #166.

Quetelet, A., and J. G. Garnier, *Correspondance mathématique et physique*, 11 vols (1825–9, 1832, 1835–9), Rare Books Collection, Boston Public Library.

Quetelet, A., *Quetelet Varia (QV)*, 4 vols, Bibliothèque Royale de Belgique, Brussels.

Raoul, L.-V., *L'Anti-Hugo* (Brussels: Kiessling, 1840).

Raeff, M., *The Well-Ordered Police State: Social and Institutional Change through Law in the Germanies and Russia, 1600–1800* (New Haven, CT: Yale University Press, 1983).

Randeraad, N., 'The International Statistical Congress (1853–1876): Knowledge Transfers and their Limits', *European History Quarterly*, 41 (2011), pp. 50–65.

Reedy, W. J., 'Language, Counter-Revolution and the "Two Cultures": Bonald's Traditionalist Scientism', *Journal of the History of Ideas*, 44 (1983), pp. 579–97.

Reichesberg, N., *Der berühmte Statistiker Adolf Quételet, sein Leben und sein Wirken* (Bern: Buchdruckerei Stämpfli, 1896).

Reiffenberg, F. B. de, 'De la direction actuellement nécessaire aux etudes philosophique', in A. Baron (ed.), *Collection d'opuscules philosophique et littéraires* (Brussels: Wahlen, 1840), pp. 65–135.

Saint-Beuve, C. A., *Portraits de femmes* (Paris: Garnier, 1882).

Sarton, G., 'Preface to Volume XXIII of *Isis* (Quetelet)', *Isis*, 23 (1935), pp. 6–24.

Schaffer, S., 'Babbage's Intelligence: Calculating Engines and the Factory System', *Critical Inquiry*, 21 (1994), pp. 203–27.

—, 'Experimenters' Techniques, Dyers' Hands, and the Electronic Planetarium' *Isis*, 98 (1997), pp. 456–83.

—, 'Astronomers Mark Time: Discipline and the Personal Equation', *Science in Context*, 2 (1998), pp. 5–29.

Schneider, U. J., 'Eclecticism Rediscovered', *Journal of the History of Ideas*, 59 (1998), pp. 173–82.

Schweber, S. S., 'The Origin of the "Origin" Revisited', *Journal for the History of Biology*, 10 (1977), pp. 229–316.

Scott, J. C., *Seeing like a State: How Certain Schemes to Improve the Human Condition Have Failed* (New Haven, CT: Yale University Press, 1998).

Seneta, E., 'Regularity and Free Will: L. A. J. Quetelet and P. A. Nekrasov', *International Statistical Review / Revue internationale de statistique*, 71 (2003), pp. 319–34.

Sepper, D., *Goethe contra Newton: Polemics and the Project for a New Science of Color* (Cambridge: Cambridge University Press, 1988).

—, 'Goethe, Colour and the Science of Seeing', in A. Cunningham and N. Jardine (eds), *Romanticism and the Sciences* (Cambridge: Cambridge University Press, 1990), 189–98.

Shapere, D., 'External and Internal Factors in the Development of Science', *Science & Technology Studies*, 4 (1986), pp. 1–9.

Shapin, S., 'Discipline and Bounding: The History and Sociology of Science as Seen through the Externalism–Internalism Debate', *History of Science*, 30 (1992), pp. 333–69.

—, 'The Man of Science', in K. Park, R. Porter and L. Daston (eds), *The Cambridge History of Science*, 7 vols, vol. 3: *Early Modern Science* (Cambridge: Cambridge University Press, 2006), pp. 179–91.

Shapin, S., and S. Schaffer, *Leviathan and the Air-Pump: Hobbes, Boyle, and the Experimental Life* (Princeton, NJ: Princeton University Press, 1985).

Siegel, L. J., *Criminology: The Core*, 3rd edn (2002; Belmont, CA: Wadsworth, 2008).

Simon, W. M., 'The Two Cultures in Nineteenth-Century France: Victor Cousin and Auguste Comte', *Journal of the History of Ideas*, 26 (1965), pp. 45–58.

Skinner, Q., 'Meaning and Understanding in the History of Ideas', *History and Theory*, 8 (1969), pp. 3–53.

Smith, R. W., 'A National Observatory Transformed: Greenwich in the Nineteenth Century', *Journal for the History of Astronomy*, 22 (1991), pp. 5–29.

—, 'Remaking Astronomy: Instruments and Practice in the Nineteenth and Twentieth Centuries', in M. J. Nye (ed.), *Cambridge History of Science*, 7 vols, vol. 5: *The Modern Physical and Mathematical Science* (Cambridge: Cambridge University Press, 2002), pp. 154–73.

Smith, R., *Free Will and the Human Sciences in Britain, 1870–1910* (London: Pickering & Chatto, 2013).

Snow, C. P., *The Two Cultures and the Scientific Revolution* (Cambridge: Cambridge University Press, 1959).

Sperber, J., *Karl Marx: A Nineteenth-Century Life* (New York: W. W. Norton & Co., 2013).

Staël, G. de, *De l'Allemagne* (1813; Paris: Garnier, 1866).

—, *De la littérature considérée dans ses rapports avec les institutions sociales* (1800; Geneva: Droz, 1959).

Stanley, M., 'The Pointsman: Maxwell's Demon, Victorian Free Will, and the Boundaries of Science', *Journal for the History of Ideas*, 69 (2008), pp. 467–91.

Stichweh, R., *Zur Entstehung des modernen Systems wissenschaftlicher Disziplinen: Physik in Deutschland, 1740–1890* (Frankfurt: Suhrkamp, 1984).

Stigler, S. M., *The History of Statistics: The Measurement of Uncertainty before 1900* (Cambridge, MA: Harvard University Press, 1986).

—, 'Adolphe Quetelet: Statistician, Scientist, Builder of Intellectual Institutions', *Actualité et universalité de la pensée scientifique d'Adolphe Quetelet: Actes du colloque organisé à l'occasion du bicentenaire de sa naissance* (Brussels: Académie royale de Belgique, 1997), pp. 47–61.

Talmon, J. L., *The Origins of Totalitarian Democracy* (New York: Praeger, 1961).

Tocqueville, A. de, *The Old Regime and the Revolution*, trans. S. Gilbert (French, 1856; Chicago, IL: University of Chicago Press, 1998).

Todhunter, I., *A History of the Mathematical Theory of Probability from the Time of Pascal to that of Laplace* (Cambridge: Macmillan & Co., 1865).

Turner, S. P., *The Search for a Methodology of Social Science* (Boston, MA: D. Reidel Pub. Co., 1986).

Van de Weyer, S., 'Discours sur l'histoire de la philosophie', in A. Baron (ed.), *Collection d'opuscules philosophique et littéraires* (Brussels: Wahlen, 1840), pp. 1–64.

Van Meenen, P. F., 'De l'influence de libre arbitre de l'homme sur les faits sociaux', *Mémoires de l'Académie royale des sciences, des lettres et des beaux-arts de Belgique*, 21 (1848), pp. 93–112.

Veblen, T., 'The Technicians and the Revolution', in M. Lerner (ed.), *The Portable Veblen* (1921; New York: Viking, 1948), pp. 438–65.

Vermeren, P. *Victor Cousin: Le Jeu de la philosophie et de l'état* (Paris: Harmattan, 1995).

Weber, M., 'Science as a Vocation', in M. Weber, *From Max Weber: Essays in Sociology*, ed. and trans. H. H. Gerth and C. W. Mills (German, 1918; New York and Oxford: Oxford University Press, 1949).

—, *Economy and Society: An Outline of Interpretive Sociology*, ed. and trans. G. Roth and C. Wittich (Berkeley, CA: University of California Press, 1978).

Wulf, M. de, *Histoire de la philosophie en Belgique* (Brussels: Dewitt, 1910).

Latour, B., and S. Woolgar, *Laboratory Life: The Construction of Social Facts*, 2nd edn (1986; Princeton, NJ: Princeton University Press, 1986).

Wright, A., *Cataloging the World: Paul Otlet and the Birth of the Information Age* (Oxford: Oxford University Press, 2014).

Wokler, R., 'The Enlightenment and the French Revolutionary Birth Pangs of Modernity' in J. Heilbron, L. Magnusson and B. Wittrock (eds), *The Rise of the Social Sciences and the Formation of Modernity: Conceptual Change in Context, 1750–1850* (Dodrecht: Kluwer Academic Publishers, 1997), pp. 35–76.

Yeo, R., *Defining Science: William Whewell, Natural Knowledge and Public Debate in Early Victorian Britain* (Cambridge: Cambridge University Press, 1993).

Young, E., *Labor in Europe and America* (Philadelphia, PA: S. A. George and Co., 1875).

Žižek, S., *Did Somebody Say Totalitarianism?* (London: Verso, 2001).

NOTES

Introduction: Two Average Men

1. A. Quetelet, *Sciences mathématique et physique chez les belges, au commencement du XIXe siècle* (Brussels: Buggenhoudt, 1866), p. 5. All French translation are my own unless otherwise noted.
2. Quetelet, *Sciences mathématique et physique chez les belges*, p. 7.
3. Quetelet, *Sciences mathématique et physique chez les belges*, p. 1.
4. The term 'scientist' was coined by Quetelet's correspondent and frequent critic William Whewell. The English term, however, did not gain much traction during Quetelet's lifetime. Therefore I will primarily refer to Quetelet and his colleagues as 'men of science', 'science workers' or what they called themselves: *savants*. See R. Yeo, *Defining Science: William Whewell, Natural Knowledge and Public Debate in Early Victorian Britain* (Cambridge: Cambridge University Press, 1993), p. 5. For a development of the term, see S. Shapin, 'The Man of Science', in K. Park, R. Porter and L. Daston (eds), *The Cambridge History of Science: Early Modern Science* (Cambridge: Cambridge University Press, 2006), pp. 179–91, and J. Morrell and A. Thackray, *Gentleman of Science: Early Years of the British Association of the Advancement of Science* (Oxford: Oxford University Press, 1981).
5. A. Quetelet, *Sur l'homme et le développement de ses facultés*, 2 vols (Brussels: Bachelier, 1835), vol. 2, pp. 279–80.
6. See below, in the section entitled 'After Bielefeld', for a full review of this literature.
7. This is not to deny the relevance of family and home life on the scientific experience. Browne, for example, has provided convincing evidence for the importance of Darwin's family in the development of his ideas. J. Browne, *Charles Darwin: The Power of Place* (New York: Alfred A. Knopf, 2003).
8. C. Dickens, *Hard Times For these Times*. (London: Bradbury & Evans, 1854), p. 4.
9. The brief polity of the United Kingdom of the Netherlands, created out of the wreckage of the Napoleonic War, lasted from 1815 until 1830. Because of the rather cumbersome nature of the name, I have mostly referred to the region where Quetelet worked as 'Belgium' or the Low Countries for the pre-1830 period, except in cases where direct reference is being made to the state.
10. A. Desrosières, *The Politics of Large Numbers: A History of Statistical Reasoning*, trans. C. Nash (Cambridge, MA: Harvard University Press, 1998), p. 79.
11. P. Buck, 'From Celestial Mechanics to Social Physics: Discontinuity in the Development of the Sciences in the Early Nineteenth Century', in H. N. Janke and M. Otte

12. (eds), *Epistemological and Social Problems of the Sciences in the Early Nineteenth Century* (Dordrecht: D. Reidel, 1981), pp. 19–33, on p. 21.
12. For a recent example for the enthusiasm for the quantitative mode of reasoning, see A. Pentland, *Social Physics: How Good Ideas Spread – Lessons from a New Science* (New York: Penguin Press, 2014). That this book largely ignores the history of its own ideas can be found in the second half of the subtitle. Like many books from the past decade which embrace a quantitative account of human behavior, Pentland fails to mention Quetelet or Comte, the two men who invented the term that serves as the title of the book.
13. For works that treat the rise (and fall) of this form of scientific thinking see E. Barkan, *The Retreat of Scientific Racism* (Cambridge: Cambridge University Press, 1993), D. J. Kevles, *In the Name of Eugenics: Genetics and the Uses of Human Heredity* (New York: Knopf, 1985) and R. Proctor, *Racial Hygiene: Medicine under the Nazis* (Cambridge: Cambridge University Press, 1988).
14. F. Manuel, 'From Equality to Organicism', *Journal of the History of Ideas*, 17 (1956), pp. 54–69, on p. 54. This idea is expanded upon in F. Manuel, *The New World of Henri Saint-Simon* (Cambridge, MA: Harvard University Press, 1956) and F. Manuel, *The Prophets of Paris* (Cambridge, MA: Harvard University Press, 1962).
15. L. Goldman, 'The Origins of British "Social Science": Political Economy, Natural Science and Statistics, 1830–1835', *Historical Journal*, 26 (1983), pp. 587–614, on p. 610.
16. T. F. Gieryn, 'Boundary-Work and the Demarcation of Science from Non-Science: Strains and Interests in Professional Ideologies of Scientists', *American Sociological Review*, 48 (1983), pp. 781–95.
17. Some of the best work on the interpenetration of science and politics from the era can be found in C. C. Gillspie, *Science and Polity in France: The Revolutionary and Napoleonic Years* (Princeton, NJ: Princeton University Press, 2004), H. W. Paul, *From Knowledge to Power: The Rise of the Science Empire in France, 1860–1939* (Cambridge: Cambridge University Press, 1995) and D. Mackenzie, *Statistics in Britain, 1865–1930* (Edinburgh: Edinburgh University Press, 1981).
18. [C. Morgan], 'Review of Quetelet's *On Man*', *Athenaeum*, 406 (1835), pp. 593–5, 611–13, 658–61, on p. 661.
19. G. Sarton, 'Preface to Volume XXIII of *Isis* (Quetelet)', *Isis*, 23 (1935), pp. 6–24, on p. 6.
20. A colleague once expressed this problem when she told me that all historians of science either 'love' or 'hate' their subjects. For the record, I do not feel either emotion toward Quetelet. This should not be taken to mean that Quetelet has not provided certain inspirations to historians of science. Stigler, for one, has shown a passionate engagement in the statistical work of Quetelet, collecting many of Quetelet's writings and deriving joy from replicating (and at times correcting) the Belgian's calculations. S. M. Stigler, 'Adolphe Quetelet: Statistician, Scientist, Builder of Intellectual Institutions', *Actualité et universalité de la pensée scientifique d'Adolphe Quetelet: Actes du colloque organisé à l'occasion du bicentenaire de sa naissance* (Brussels: Académie royale de Belgique, 1997), pp. 47–61.
21. The move toward practice was led in part by the discipline of science studies, the sometime-contentious cousin of the history of science. Influential works include B. Latour, *Science in Action: How to Follow Scientists and Engineers through Society* (Cambridge, MA: Harvard University Press, 1987) and B. Latour and S. Woolgar, *Laboratory Life: The Construction of Social Facts*, 2nd edn (Princeton, NJ: Princeton University Press, 1986). For more on how these two disciplines have influenced one another, see P. Dear and S. Jasanoff, 'Dismantling Boundaries in Science and Technology Studies', *Isis*, 101 (2010), pp. 759–74.

22. See I. B. Cohen, *Revolution in Science* (Cambridge, MA: Belknap, 1985), pp. 91–101. To muddle things further, the term 'Second Scientific Revolution' has also been used to cover the *entire* period from 1800 until 1950. S. G. Brush, *The History of Modern Science: A Guide to the Second Scientific Revolution* (Ames, IA: Iowa State University Press, 1988).
23. T. Kuhn, 'The Function of Measurement in Modern Physical Sciences', *Isis*, 52 (1961), pp. 161–93, on p. 188. Emphases are in the original.
24. Hahn was the first to use the phrase 'second scientific revolution' in relation to the institutional changes in scientific production. R. Hahn, *The Anatomy of a Scientific Institution: The Paris Academy of Sciences, 1666–1803* (Berkeley, CA: University of California Press, 1971), p. 275.
25. This argument dates to at least J. T. Merz, *A History of European Thought in the Nineteenth Century*, 2 vols (Edinburgh: Blackwood, 1912). More recent has been the influence of R. Stichweh, *Zur Entstehung des modernen Systems wissenschaftlicher Disziplinen: Physik in Deutschland, 1740–1890* (Frankfurt: Suhrkamp, 1984), J. Golinski, *Making Natural Knowledge: Constructivism and the History of Science* (Cambridge: Cambridge University Press, 1998), p. 67, D. Outram, 'Politics and Vocation: French Science, 1793–1830', *British Journal for the History of Science*, 13 (1980), pp. 27–43 and R. Yeo, *Defining Science: William Whewell, Natural Knowledge and Public Debate in Early Victorian Britain* (Cambridge: Cambridge University Press, 1993). Such studies all deal, directly or indirectly with M. Weber, 'Science as a Vocation', in M. Weber, *From Max Weber: Essays in Sociology*, ed. H. H. Gerth and C. W. Mills (German, 1918; New York: Oxford University Press, 1949).
26. Terms like 'professionalization' have often been employed for some forms of scientific research, sometimes ranging over extraordinarily large geographical and chronological spans, and obviously a single term cannot do justice to a process that includes such varied personalities as Davy, Ampère, Pasteur, Laplace, Darwin, Maxwell, Fourier, Priestly, Mendel, Herschel, Humboldt and hundreds upon thousands of other Europeans and Americans who identified their research as science. For a survey of the strengths and weaknesses of this terminology, see P. Lucier, 'The Professional and the Scientist in Nineteenth-Century America', *Isis*, 100 (2009), pp. 699–732, on pp. 700–3.
27. A. Quetelet, *Sciences mathématique et physique chez les belges, au commencement du XIXe siècle* (Brussels: Buggenhoudt, 1866), p. 5, and A. Quetelet, *Histoire des sciences mathématique et physique chez les belges* (1864), new edn (Brussels: Muquart, 1871).
28. The lives of this 'new class of man' who helped provide much of the intellectual and even physical labor for nineteenth-century science have been traced in several locations, including several works by Simon Schaffer that have described the conditioning of low-level sciences workers in observatories and factories. See S. Schaffer, 'Astronomers Mark Time: Discipline and the Personal Equation', *Science in Context*, 2 (1998), pp. 5–29; 'Babbage's Intelligence: Calculating Engines and the Factory System', *Critical Inquiry*, 21 (1994), pp. 203–27 and 'Experimenters' Techniques, Dyers' Hands, and the Electronic Planetarium', *Isis*, 98 (1997), pp. 456–83. For a series of informative posts on Schaffer's efforts to connect physical labour to scientific practice, see the weblog Ether Wave Propaganda at https://etherwave.wordpress.com/category/schaffer-oeuvre/ [accessed 1 November 2014]. The average men imagined for Quetelet's social physics were not quite modeled on factory life like Schaffer's dyers and computers, however, but rather on the administrative life, a relatively less-studied phenomenon of the nineteenth century sciences. As a recent study of 'The Organizational Revolution' has claimed, bureaucracy has been overlooked as a 'fertile source of models and metaphors in sci-

ence', receiving far less attention than 'the factory and the assemble line'. Furthermore it is only in recent years that 'the bureaucrat as an observer' has begun to be studied. H. Heyck, 'The Organizational Revolution and the Human Sciences', *Isis*, 105 (2014), pp. 1–31, on p. 18, and P. Becker and W. Clark, 'Introduction', in P. Becker and W. Clark (eds), *Little Tools of Knowledge: Historical Essays on Academic and Bureaucratic Practices, Social History, Popular Culture, and Politics* (Ann Arbor, MI: University of Michigan Press, 2001), pp. 1–34, on p. 12. Quetelet's plans for a science of man may offer one of the best examples where the bureaucratic metaphor is apt. Quetelet has been described as a 'bureaucratic liberal', and one of the principal studies in the practice of probabilistic sciences has described his average men as 'statistically-minded bureaucrats'. G. Gigerenzer et al., *The Empire of Chance: How Probability Changes Science and Everyday Life* (Cambridge: Cambridge University Press, 1989), p. 43 and C. C. Gillispie, with R. Fox and I. Grattan-Guinness, *Pierre-Simon Laplace 1749–1827: A Life in Exact Science* (Princeton, NJ: Princeton University Press, 1997), p. 275.

29. The last biographies of Quetelet in English and French were published in 1908 and 1912, respectively: F. H. Hankins, *Adolphe Quetelet as Statistician* (New York: Columbia University Press, 1908), pp.1–136 and J. Lottin, *Quetelet, statisticien et sociologue* (Louvain: Institut supérieur de philosophie, 1912).

30. M. Halbwachs, *La Théorie de l'homme moyen: Essais sur Quetelet et la statistique morale* (Paris: Librairie Félix Alcan, 1913), p. 7.

31. L. Krüger, L. J. Daston and M. Heidelberger (eds), *The Probabilistic Revolution*, 2 vols, vol. 1: *Ideas in History*, and G. Gigerenzer, L. Krüger and M. S. Morgan (eds), *The Probabilistic Revolution*, 2 vols, vol. 2: *Ideas in the Sciences* (Cambridge, MA: Massachusetts Institute of Technology Press, 1987). Many of the participants also contributed to the colloquium held in 1996 at the Académie royale de Belgique to mark the bicentennial of Quetelet's birth. See below.

32. L. Daston, *Classical Probability in the Enlightenment* (Princeton, NJ: Princeton University Press, 1988), I. Hacking, *The Taming of Chance* (Cambridge: Cambridge Press, 1990), T. M. Porter, *The Rise of Statistical Thinking, 1820–1900* (Princeton, NJ: Princeton University Press, 1986) and S. Stigler, *The History of Statistics: the Measurement of Uncertainty Before 1900* (Cambridge, MA: Belknap Press, 1986).

33. P. Beirne, 'Adolphe Quetelet and the Origins of Positivist Criminology', *American Journal of Sociology*, 92 (1987), pp. 1140–69; L. Goldman, 'The Origins of British "Social Science": Political Economy, Natural Science and Statistics, 1830–1835', *Historical Journal*, 26 (1983), pp. 587–614, on p. 610; P. F. Lazerfeld, 'Notes on the History of Quantification in Sociology – Trends, Sources and Problems', *Isis*, 52 (1961), pp. 277–333 and S. S. Schweber, 'The Origins of the *Origin* Revisited', *Journal of the History of Biology*, 10 (1977), pp. 229–316.

34. *Actualité et universalité de la pensée scientifique d'Adolphe Quetelet* (Brussels: Académie royale de Belgique, 1997).

35. Stigler *did* recognize the importance of a synthesis between Quetelet's ideas and his institutional work and offered one possible idea for 'the fundamental unity in his work'. Stigler's answer, however, was that Quetelet practised a kind of 'social meteorology', rather than the more deterministic 'social physics', that united thought and practice throughout his career. While a provocative and clever idea, it had two problems, even aside from the fact that Quetelet *did* call his theory 'social physics'. The first was that Stigler interpreted 'meteorology' in our current sense as a highly probabilistic science. However, Quetelet assumed that meteorology would eventually reach the predictive

level of physics and the difference we now see between 'social meteorology' and 'social physics' would not have occurred to Quetelet. The second problem is that it did not incorporate Quetelet's own philosophy of science as a highly regulated and organized research body, but interpreted his status as 'institution builder' as merely a position of power. S. M. Stigler, 'Adolphe Quetelet: Statistician, Scientist, Builder of Intellectual Institutions', *Actualité et universalité de la pensée scientifique d'Adolphe Quetelet: Actes du colloque organisé à l'occasion du bicentenaire de sa naissance* (Brussels: Académie royale de Belgique, 1997), pp. 47–61.

36. The fragmented literature has continued since this conference. On Quetelet and the creation of family life, see B. P. Cooper, *Family Fictions and Family Facts: Harriet Martineau, Adolphe Quetelet, and the Population Question in England* (New York: Routledge, 2007). On his importance in 'contemporary discussions between mainstream and heterodox economic theorists', see B. Mosselmans, 'Adolphe Quetelet, the Average Man and the Development of Economic Methodology', *European Journal of the History of Economic Thought*, 12 (2005), pp. 562–85, on p. 577.

37. Steven Shapin reported that by the early nineties, 'there was much sighing and rolling of eyes' among historians when he brought up the topic. S. Shapin, 'Discipline and Bounding: The History and Sociology of Science as Seen through the Externalism–Internalism Debate', *History of Science*, 30 (1992), pp. 333–69, on p. 334. For other treatments of the methodological debate, see B. T. Paller, 'Naturalized Philosophy of Science, History of Science, and the Internal/External Debate', *Proceedings of the Biennial Meeting of the Philosophy of Science Association*, 1 (1986), pp. 258–68, and D. Shapere, 'External and Internal Factors in the Development of Science', *Science & Technology Studies*, 4 (1986), pp 1–9.

38. Q. Skinner, 'Meaning and Understanding in the History of Ideas', *History and Theory*, 8 (1969), pp. 3–53.

39. Skinner, 'Meaning and Understanding in the History of Ideas', p. 11.

40. Skinner himself may have been guilty of this charge. One critic claimed that Skinner contextualized so much in his work on Machiavelli that the great Florentine was reduced from a 'mountain on the landscape of political theory ... to a molehill'. M. Hulling, *Citizen Machiavelli* (Princeton, NJ: Princeton University Press, 1983), p. 26.

41. For an fine example of history writing in this vein, see K. M. Baker, *Inventing the French Revolution: Essays on French Political Culture in the Eighteenth Century* (Cambridge: Cambridge University Press, 1990).

42. P. Bourdieu, 'The Peculiar History of Scientific Reason', *Sociological Forum*, 6 (1991), pp. 3–26.

43. This insight was not limited to science. See P. Bourdieu, *Les règles de l'art: Genèse et structure du champ littéraire* (Paris: Seuil, 1992).

44. The idea that much of what constituted scientific work was determined by non-scientific processes, as historians of science know well, had received attention prior to Bourdieu. Two highly influential works that have demonstrated this point are S. Shapin and S. Schaffer, *Leviathan and the Air-Pump: Hobbes, Boyle, and the Experimental Life* (Princeton, NJ: Princeton University Press, 1985) and T. Kuhn, *The Structure of Scientific Revolutions* (Chicago, IL: University of Chicago Press, 1962).

45. Bourdieu, 'The Peculiar History of Scientific Reason', p. 10.

46. R. Collins, *The Sociology of Philosophies: A Global Theory of Intellectual Change* (Cambridge, MA: Belknap Press, 1998). Popular works of intellectual history influenced by this approach include J. Buchnan, *Crowded with Genius: The Scottish Enlightenment,*

Edinburgh's Moment of the Mind (New York: Harper Collins, 2003) and L. Menand, *The Metaphysical Club* (New York: Farrar, Straus and Giroux, 2001).

47. Collins, *The Sociology of Philosophies*, p. 77. So strong was the determinism that the original title for Collins's work was *The Social Causes of Philosophies*. S. Fuchs, 'A Sociological Theory of Scientific Change', *Social Forces*, 71 (1993), pp. 933–53, on p. 950.
48. Collins, *The Sociology of Philosophies*, p. 54.
49. M. Biagioli, *Galileo, Courtier: The Practice of Science in the Culture of Absolutism* (Chicago, IL: University of Chicago Press, 1993) and S. F. Cannon, *Science in Culture: The Early Victorian Period* (New York: Science History Publications, 1978).
50. Collins, *The Sociology of Philosophies*, p. 82.
51. I am also not persuaded by the most rigid attempts to go beyond Bourdieu, such as the recent work on SIMs (scientific/intellectual movements). An example would be S. Frickel and N. Gross, 'A General Theory of Scientific/Intellectual Movements', *American Sociological Review*, 70 (2005), pp. 204–32.
52. For an excellent survey of Einstein and his relation to his time, see G. J. Holton, P. Galison and S. S. Schweber (eds), *Einstein for the 21st Century: His Legacy in Science, Art, and Modern Culture* (Princeton, NJ: Princeton University Press, 2008).
53. Bourdieu, 'The Peculiar History of Scientific Reason' p. 16.
54. R. K. and A. Reisenger (eds), *Climate Change 2007: Synthesis Report* (Geneva: International Panel on Climate Control, 2007), p. 121.

1 Life in the War: The End of Enlightenment in Belgium, 1796–1823

1. Great Britain Geographical Section of the Naval Intelligence Division, *A Manual of Belgium and the Adjoining Territories* (London: His Majesty's Stationary Office, 1918).
2. H. Pirenne, *La Nation belge* (Brussels: Guyot, 1899), p. 3. On the importance of boundaries and frontiers for the development of Belgium, see the introduction to volume 1 in Pirenne, *La Nation belge*.
3. For two accounts of the Enlightenment influence in Belgium, see H. Francotte, 'Essai historique sur la propagande des encyclopédistes français à Liège', *Mémoires couronnés et autres mémoires*, 30 (1880); and M. J. Kuntzinger, 'Essai historique sur la propagande des encyclopédistes français en Belgique dans la seconde moitié de huitième siècle', *Mémoires couronnes et autres mémoires*, 30 (1880).
4. Pirenne, *La Nation belge*, p. xi. Though it may seem too much a simplification to reduce the French and German intellectual movements of the time to positivism and Romanticism, respectively, or even to something like the sciences and the arts, the nuances found by later interpreters were not always evident at the time. Though in retrospect partisans on each side may have made too much of the war between the sciences and the arts, there was a strong belief at the time of a divide between these forms of knowledge, real or imagined.
5. J. Heilbron, *The Rise of Social Theory* (Cambridge: Polity Press, 1995), pp. 113–30.
6. Heilbron, *The Rise of Social Theory*, pp. 14–15.
7. L. de Bonald, 'Sur la guerre des sciences et lettres', in L. de Bonald, *Oeuvres de M. Bonald: Mélanges littéraires, politiques et philosophiques*, 8 vols (Brussels: Gérant, 1845), vol. 7, pp. 459–64.
8. L. de Bonald, 'Des sciences, des lettres et des arts', in L. de Bonald, *Oeuvres de M.*

Bonald: *Mélanges littéraires, politiques et philosophiques*, 8 vols (1807; Brussels: Gérant, 1845), vol. 7, pp. 349–80.
9. Bonald, 'Sur la guerre des sciences et lettres', p. 459.
10. While Bonald and Chateaubriand shared similar attitudes towards what they believed to be an excessive focus on empirical science, recent work has helped shed light on his more nuanced position between positivism and Romanticism. D. Klinck, *The French Counterrevolutionary Theorist Louis de Bonald (1754–1840)* (New York: P. Lang, 1996). For an excellent summary of Bonald's position in relation to the 'two cultures' see W. J. Reedy, 'Language, Counter-Revolution and The "Two Cultures": Bonald's Traditionalist Scientism', *Journal of the History of Ideas*, 44 (1983). pp. 579–97.
11. Bonald, 'Des sciences, des lettres et des arts', p. 350.
12. Bonald, 'Des sciences, des lettres et des arts', p. 351.
13. Bonald, 'Des sciences, des lettres et des arts', p. 351.
14. Bonald, 'Des sciences, des lettres et des arts', p. 352.
15. Bonald, 'Des sciences, des lettres et des arts', p. 354.
16. Bonald, 'Des sciences, des lettres et des arts', p. 363.
17. Bonald, 'Des sciences, des lettres et des arts', p. 364.
18. Bonald, 'Des sciences, des lettres et des arts', p. 355.
19. H. Grégoire, 'Education in Belgium', in J.-A. Goris (ed.), *Belgium* (Berkeley, CA: University of California Press, 1945), pp. 226–38, on p. 236.
20. H. Francotte, 'Essai historique sur la propagande des encyclopédistes français à Liège', *Mémoires couronnés et autres mémoires*, 30 (1880).
21. G. de Staël, *De l'Allemagne* (1813; Paris: Garnier, 1866), p. 13.
22. Staël, *De l'Allemagne*, p. 92.
23. J.-L. de Paepe, *Exposition documentaire présentée à la Bibliothèque Royale Albert I à l'occasion du centiare de la mort d'Adolphe Quetelet* (Brussels: Académie royale de Belgique, 1974).
24. Quetelet Archives, Académie royale de Belgique, Brussels, file #2. Hereafter abbreviated as ARB. Navigating these materials would have been impossible without the work of Liliane Wellens-De Donder. L. W.-De Donder, *Inventaire de la correspondance d'Adolphe Quetelet déposé à l'l'Académie royale de Belgique* (Brussels: L'Académie royale de Belgique, 1968).
25. A. Quetelet, 'Dominique-François-Jean Arago', *Annuaire de l'Académie royale de Belgique* (1855), contained in *Quetelet Varia* (*QV*), a collection of miscellany held at the Bibliothèque royale de Belgique. *QV*, 4 vols, vol. 4, #127/8, p. 159. (Note: This 'Notice' is unnumbered in the *QV* but is placed between numbers 127 and 128 in the fourth volume.) For the influence of Arago on Quetelet, see below, Chapter Four.
26. Kuhn of course makes such professionalization a necessary cause for the structure of his revolutions to be possible. T. Kuhn, *The Structure of Scientific Revolutions* (Chicago, IL: University of Chicago Press, 1962). See also, T. F. Gieryn, 'Boundary-Work and the Demarcation of Science from Non-Science: Strains and Interests in Professional Ideologies of Scientists', *American Sociological Review*, 48 (1983), pp. 781–95.
27. The explosion of works on the history of probability and statistics has been remarkable in the past thirty years. See the introduction for details and references. Yet Quetelet did not necessarily have the same opportunities as his influences.
28. B. Pascal, 'Préface sur le traité du vide', in B. Pascal, *Pascal: Oeuvres complètes*, ed. L. Lafuma (1651; Paris: Éditions du Seuil, 1963), pp. 230–2, on p. 230.
29. B. Pascal, 'Discours sur les passions de l'amour', in B. Pascal, *Pascal: Oeuvres complètes*,

ed. L. Lafuma (1652–3; Paris: Éditions du Seuil, 1963), pp. 285–9, on p. 286. Pascal's authorship of the *Discours* remains in doubt, but it is included in most editions of his work and is seen on the whole to be at least representative of his ideas. Lafuma provides a good summary of the question in his introduction on p. 285, though he questions the 1652 dating.

30. E. Mailly, 'Essai sur la vie et les ouvrages de Lambert-Adolphe-Jaques Quetelet', *Annuaire de l'Académie royale de Belgique* (1875), pp. 108–297, on p. 116.
31. E. Cassirer, *The Philosophy of the Enlightenment*, trans. F. C. A. Koelln and J. P. Pettegrove (German, 1932; Princeton, NJ: Princeton University Press, 1951), p. 35.
32. T. L. Hankins, *Jean d'Alembert: Science and the Enlightenment* (Oxford: Clarendon Press, 1970), p. 71.
33. Hankins, *Jean d'Alembert: Science and the Enlightenment*, p. 132.
34. J. la R. d'Alembert, Alembert, J. la R. d', 'Essai sur la société des gens de lettres et des grands', in J. la R. d'Alembert, *Oeuvres Complètes de d'Alembert* (1753; Geneva: Slatkine, 1967), pp. 335–72, on p. 341.
35. K. M. Baker, *Condorcet: From Natural Philosophy to Social Mathematics* (Chicago, IL: University of Chicago Press, 1975), p. 372.
36. J. Lottin, *Quetelet, statisticien et sociologue* (Louvain: Institut supérieur de philosophie, 1912), pp. 371–8.
37. F. Manuel, *The Prophets of Paris* (Cambridge, MA: Harvard University Press, 1962).
38. Baker, *Condorcet*, p. 10.
39. This is also the picture to emerge from the first volume of Gillispie's work on the French sciences, C. C. Gillispie, *Science and Polity in France: The End of the Old Regime* (Princeton, NJ: Princeton University Press, 1980).
40. P.-S. de Laplace, *A Philosophical Essay on Probabilities*, trans. F. W. Truscott and F. L. Emory (French, 1814; New York: Dover Publications, 1951), pp. 107–8.
41. P.-S. Laplace, *Essai philosophique sur les probabilités*, (1814; Brussels: Culture and Civilization, 1967).
42. C. C. Gillispie, with R. Fox and I. Grattan-Guinness, *Pierre-Simon Laplace 1749–1827: A Life in Exact Science* (Princeton, NJ: Princeton University Press, 1997), p. 183.
43. For the full story see C. C. Gillispie, *Science and Polity in France: The Revolutionary Years* (Princeton, NJ: Princeton University Press, 2004).
44. F.-R. de Chateaubriand, *Mémoires d'outre-tombe*, 4 vols (1849–50; Paris: Garnier, 1998), vol. 2, p. 647.
45. A. Quetelet, 'Essai sur la romance', *Annales Belgique*, 8 (1823), pp. 223–32, 295–311.
46. F.-R. de Chateaubriand, 'Essai sur les révolutions', in F.-R. de Chateaubriand, *Essai sur les révolutions, Génie du christianisme*, ed. M. Regard (1797; Paris: Gallimard, 1978), p. 120.
47. Chateaubriand, 'Essai sur les révolutions', p. 371.
48. F.-R. de Chateaubriand, 'Génie du christianisme', in F.-R. de Chateaubriand, *Essai sur les Révolutions; Génie du Christianisme*, ed. M. Regard (1802; Paris: Gallimard, 1978), p. 468.
49. Chateaubriand, 'Génie du christianisme', p. 812.
50. Chateaubriand, 'Génie du christianisme', p. 812.
51. Chateaubriand, 'Génie du christianisme', p. 814.
52. Chateaubriand, 'Génie du christianisme', p. 817.
53. C.A. Saint-Beuve, *Portraits de femmes* (Paris: Garnier, 1882), pp. 83, 115.
54. G. de Staël, *De la littérature considérée dans ses rapports avec les institutions sociales* (1800; Geneva: Droz, 1959), p. 13.
55. Staël, *De la littérature considérée dans ses rapports avec les institutions sociales*, p. 14.

56. G. de Staël, *De l'Allemagne* (1813; Paris: Garnier, 1866), p. 92.
57. Staël, *De l'Allemagne*, pp. 92–3.
58. L. Daston, *Classical Probability in the Enlightenment* (Princeton, NJ: Princeton University Press, 1988).
59. F.-R. de Chateaubriand, 'Lettre au C. Fontanes sur la seconde édition de l'ouvrage de Madame de Staël', *Mercure de France*, 3 (1801), pp. 14–38.
60. A. de Lamartine, *Des destinées de la poésie* (Paris: Gosselin, 1834), p. 3.
61. Lamartine, *Des destinées de la poésie*, p. 5.
62. Lamartine, *Des destinées de la poésie*, p. 13.
63. The best recent account of Cousin can be found in C. Bolus-Reichert, *The Age of Eclecticism* (Columbus, OH: Ohio State University Press, 2009).
64. M. de Wulf, *Histoire de la philosophie en Belgique* (Brussels: Dewitt, 1910), p. 266.
65. Quoted in de Wulf, *Histoire de la philosophie en Belgique*, p. 279.
66. A. Baron, 'Preface', in A. Baron (ed.), *Collection d'opuscules philosophique et littéraires* (Brussels: Wahlen, 1840), pp. iii–x, on p. v.
67. F. B. de Reiffenberg, 'De la direction actuellement nécessaire aux études philosophique', in A. Baron (ed.), *Collection d'opuscules philosophique et littéraires*, pp. 65–135, on p. 68.
68. Reiffenberg, 'De la direction actuellement nécessaire aux études philosophique', p. 73.
69. Reiffenberg, 'De la direction actuellement nécessaire aux études philosophique', p. 68.
70. Reiffenberg, 'De la direction actuellement nécessaire aux études philosophique', p. 76.
71. Reiffenberg, 'De la direction actuellement nécessaire aux études philosophique', p. 104.
72. On the contrasting approaches of Quetelet and Reiffenberg towards the embrace of 'industrialism' by the United Kingdom of the Netherlands, see below.
73. S. Van de Weyer, 'Discours sur l'histoire de la philosophie', in A. Baron (ed.), *Collection d'opuscules philosophique et littéraires* (Brussels: Wahlen, 1840), pp. 1–64, on p. 5.
74. Van de Weyer, 'Discours sur l'histoire de la philosophie', p. 51.
75. V. Cousin, *Fragments de philosophie cartésienne* (Paris: Didier, 1852), p. vi.
76. V. Cousin, *Fragments de philosophie cartésienne*, p. vi.
77. P. Vermeren, *Victor Cousin: Le Jeu de la philosophie et de l'état* (Paris: Harmattan, 1995) and U. J. Schneider, 'Eclecticism Rediscovered', *Journal of the History of Ideas*, 59 (1998), pp. 173–82.
78. V. Cousin, *Cours de philosophie, professé à la faculté des lettres pendant l'année 1818 ...* (Paris: Hachette, 1836), p. 224.
79. W. M. Simon, 'The Two Cultures in Nineteenth-Century France: Victor Cousin and Auguste Comte', *Journal of the History of Ideas*, 26 (1965), pp. 45–58. In his article, Simon specifically acknowledges his debt to Snow's classic formulation of a twentieth-century academic war of the arts and sciences. C. P. Snow, *The Two Cultures and the Scientific Revolution* (Cambridge: Cambridge University Press, 1959).
80. On Quetelet and positivism, see T. M. Porter, 'Was Quetelet a Positivist?', *Actualité et universalité de la pensée scientifique d'Adolphe Quetelet* (Brussels: Académie royale de Belgique, 1996), pp. 199–209.
81. See Chapter Three.

2 Casualties of War: Quetelet and Friends in Ghent and Brussels, 1815–23

1. B. J. Becker, *Unravelling Starlight: William and Margaret Huggins and the Rise of the New Astronomy* (Cambridge: Cambridge University Press, 2011), p. 82.
2. J. Lottin, *Quetelet, statisticien et sociologue* (Louvain: Institut supérieur de philosophie, 1912), pp. 8–10 and F. H. Hankins, *Adolphe Quetelet as Statistician* (New York: Columbia University Press, 1908), pp. 1–136, on pp. 8–12.
3. P. Bergmans, 'Quetelet poète', *Ciel et terre*, 45 (1929), pp. 12–16, on p. 12.
4. E. Mailly, 'Essai sur la vie et les ouvrages de Lambert-Adolphe-Jaques Quetelet', *Annuaire de l'Académie royale de Belgique* (1875), pp. 108–297, on p. 110.
5. P. Lévy, 'Quetelet, le poète et le statisticien', in *Adolphe Quetelet, 1796–1874: Recueil des travaux et contributions présentés en 1974 en hommage à son rôle de statisticien* (Brussels: Académie royale de Belgique, 1974), pp. 43–56, on p. 44. Similarly, the idea that in Quetelet's project 'the artist becomes the near double of the scientist' seems to be a considerable projection of twentieth-century notions of authorship onto social physics. B. P. Cooper and M. S. Murphy, 'The Death of the Author at the Birth of Social Science: The Cases of Harriet Marineau and Adolphe Quetelet', *Studies in History and Philosophy of Science Part A*, 31 (2000), pp. 1–36, on p. 19.
6. J.-L. de Paepe, *Exposition documentaire présentée à la Bibliothèque Royale Albert I à l'occasion du centiare de la mort d'Adolphe Quetelet* (Brussels: Académie royale de Belgique, 1974), p. 4.
7. The list: French, English, Dutch, Italian, Spanish, Portuguese and German.
8. Quetelet Archives, Académie royale de Belgique, Brussels (ARB), file #196.
9. Mailly, 'Essai sur la vie', p. 111.
10. A. Quetelet and G. Dandelin, 'Jean-Second opera: En une acte et en prose' (1816), p. 3, ARB #166. Wherever possible, I have included the dates for the operas and poems.
11. Quetelet and Dandelin, 'Jean-Second opera: En une acte et un prose', p. 4.
12. A. Quetelet, 'Moschur' (1817), ARB #169.
13. Just a decade later Quetelet would be in regular contact with the 'Storm King' James Espy, who had high hopes for predicting and inducing storms. Quetelet was included as a recipient of Espy's 'To the Friends of Science' (n.p.: 1845). For an account of the most significant efforts of Espy to control the weather, see L. M. Morehead, *A Few Incidents in the Life of Professor James P.Espy* (Cincinnati, OH: R. Clarke & Co., 1888).
14. Quetelet, 'Moschur'.
15. A. Quetelet, 'M. Dièse ou l'auteur dans l'embarras' (1819), ARB #179.
16. A. Quetelet, 'Épître à Dandelin' (1819), ARB #193.
17. Quetelet, 'M. Dièse'.
18. Quetelet, 'M. Dièse'.
19. Quetelet, 'M. Dièse'.
20. For a future mathematics professor and editor of a scientific journal entitled *Correspondance mathématique et physique*, Quetelet's thesis on conic sections appears surprisingly as an outlier in his early bibliography. Though he would publish close to a dozen papers on theoretical mathematics during the 1820s, at the time, his *Dissertatio mathematica inauguralis de quibusdam locis geometrics nec non de curva focali* was an isolated work more important for Quetelet's future career then a reflection of his interests. Inspired by Pascal's work on the interaction of abstract geometric shapes, Quetelet sought to determine the series of curves that were described from the intersection of a cone and

a flat plane (i.e., a conic section). Quetelet's early work on curves was even (unjustly it turned out) hailed by his friend Louis-Vincent Raoul to be as important as the work of Pascal. For Raoul, Quetelet's 'discovery of a curve … [was] sufficient to make the reputation of more than one great geometer, and one will not be able to separate the name of his cycloid from that of Pascal's'. Quoted in Mailly, 'Essai sur la vie', p. 115. Mailly strongly disagreed with this assessment of his mentor's early work. Though the resulting curves were not quite as revolutionary as those of Pascal, the work was sufficient enough to impress his advisor J. G. Garnier and secure Quetelet a position at the Athénée de Bruxelles. The most important ramifications for this effort, however, would not be apparent until 1823, when Garnier asked his former student to help him co-edit *Correspondance mathématique et physique*.

21. Quetelet's collected poetry is contained in the Quetelet Archives, Académie royale de Belgique, Brussels (ARB), file #183, at the Académie royale. Though there are only thirty-one poems contained, there are listings for another seven that have subsequently been crossed out, in addition to notes on another possible half dozen. The work is divided into four categories: '*Elégies*', 'Romances and *fablier*', '*Epître*' and 'Miscellaneous'. No pagination in included for the poems, and only occasional dates are listed.
22. E. Mailly, 'Essai sur la vie et les ouvrages de Lambert-Adolphe-Jaques Quetelet', *Annuaire de l'Académie royale de Belgique* (1875), pp. 108–297, on p. 113.
23. A. Quetelet, 'La Poète mourant à son lampe' (1817), ARB #183.
24. A. Quetelet, 'La Comtesse Ide', 'Le Scalde et Lysis', 'Épître à mon sœur', ARB #183.
25. A. Quetelet, 'L'Illusion', ARB #183.
26. A. Quetelet, 'À mon ami de Reiffenberg', ARB #183.
27. Quetelet, 'Épître à Dandelin', ARB #183.
28. Quetelet, 'Épître à Dandelin'.
29. Mailly writes that it was in the style of Horace, 'the favorite poet of Quetelet's'. Mailly, 'Essai sur la vie', p. 122.
30. A. Quetelet, 'Épître à M. Tollens', ARB #183.
31. A. Quetelet, 'La Poète et la Raison', ARB #183. For d'Alembert's view, see Chapter Two.
32. A. Quetelet, *Notice sur le colonel G-P Dandelin* (Brussels: Hayez, 1848), p. 4. As permanent secretary of the Académie royale de Belgique, Quetelet wrote dozens of notices for the passing of fellow members. Though Reiffenberg achieved significant stature and received several notices at the time of his death, most of the available biographical information on Raoul, Dandelin and Garnier comes from Quetelet.
33. Quetelet, *Notice sur le colonel G-P Dandelin*, p. 5. In 1830 Dandelin would again take up his military post during the Belgian revolution, as Quetelet writes, 'for the third time in his life abandoning the pen for the sword'.
34. Quetelet, *Notice sur le colonel G-P Dandelin*, p. 22.
35. E. Mailly, 'Essai sur la vie et les ouvrages de Lambert-Adolphe-Jaques Quetelet', *Annuaire de l'Académie royale de Belgique* (1875), pp. 108–297, on p. 15.
36. Quetelet, *Notice sur le colonel G-P Dandelin*, p. 22.
37. On Dandelin's eventual struggles, see K. Donnelly, 'The Other Average Man: Science Workers in Quetelet's Belgium', *History of Science*, 52 (2014), pp. 401–28.
38. Quetelet, *Notice sur le colonel G-P Dandelin*, p. 18.
39. Quetelet, *Notice sur le colonel G-P Dandelin*, p. 16.
40. Quetelet, *Notice sur le colonel G-P Dandelin*, p. 22.
41. Quetelet, *Notice sur le colonel G-P Dandelin*, p. 9.
42. Mailly, 'Essai sur la vie', p. 129.

43. Mailly, 'Essai sur la vie', p. 113.
44. H. Grégoire, 'Education in Belgium', in J.-A. Goris (ed.), *Belgium* (Berkeley, CA: University of California Press, 1945), pp. 226–38.
45. A. Quetelet, *Notice sur Louis-Vincent Raoul* (Brussels: Hayez, 1849), p. 21.
46. Quetelet Archives, Académie royale de Belgique, Brussels (ARB), file #2085.
47. Quetelet, *Notice sur Louis-Vincent Raoul*, p. 22.
48. Quetelet, *Notice sur Louis-Vincent Raoul*, p. 6.
49. Mailly, 'Essai sur la vie', p. 129.
50. Mailly, 'Essai sur la vie', p. 129.
51. L.-V. Raoul, *L'Anti-Hugo* (Brussels: Kiessling, 1840), p. xi.
52. L.-V. Raoul, *L'Anti-Hugo*, p. xviii.
53. ARB #2085.
54. For a summary of the activities of Marx and Engels in Brussels, which includes a nice survey of the city's politics, see J. Sperber, *Karl Marx: A Nineteenth-Century Life* (New York: W. W. Norton & Co., 2013), pp. 153–64.
55. Y. Cassis, *Capitals of Capital: The Rise and Fall of International Financial Centres 1780–2009*, new edn (2006; Cambridge: Cambridge University Press, 2010), p. 33.
56. Mailly, 'Essai sur la vie', p. 121.
57. A. Quetelet, *Notice sur Frédéric-August-Ferdinand-Thomas Baron de Reiffenberg* (Brussels: Hayez, 1852).
58. Mailly, 'Essai sur la vie', p. 119.
59. C. Nieuport, *Essai sur la théorie du raisonnent* (Brussels: Lemaire, 1805), p. 297.
60. J.-G.-A. Luthereau, *Notice sur M. le Baron de Reiffenberg* (Brussels: Galarie St.-Hubert, 1850), p. 4.
61. Quetelet, *Notice sur Frédéric-August-Ferdinand-Thomas Baron de Reiffenberg*, p. 8.
62. Quetelet, *Notice sur Frédéric-August-Ferdinand-Thomas Baron de Reiffenberg*, p. 9.
63. Quetelet, *Notice sur Frédéric-August-Ferdinand-Thomas Baron de Reiffenberg*, p. 45.
64. Quetelet, *Notice sur Frédéric-August-Ferdinand-Thomas Baron de Reiffenberg*, p. 45.

3 Stoking the Sacred Fire: The Administration of Observation in the United Kingdom of the Netherlands, 1822–30

1. The relationship between science and state has been well documented, especially in France. Though accounts of state support of science date back to the earliest histories of the British Royal Society and the French Academy, the work of historians in the past decades have explained how the intellectual content of science itself has been influenced by government largesse. See especially the work of Gillispie. In addition to the references noted above in Chapter One, also see C. C. Gillispie, 'Probability and Politics', *Proceedings of the American Philosophical Society*, 116 (1972), pp. 1–20. Also for France, see M. Crosland, 'A Science Empire in Napoleonic France', *History of Science*, 44 (2006), pp. 29–48, D. Outram, *Georges Cuvier: Vocation, Science, and Authority in Post-Revolutionary France* (Manchester: Manchester University Press, 1984), D. Outram, 'Politics and Vocation: French Science, 1793–1830', *British Journal for the History of Science*, 13 (1980), pp. 27–43. For the relationship between statistics and probability to states, in addition to Gillispie see: J. L. Heilbron, T. Frängsmyr and R. E. Rider (eds), *The Quantifying Spirit in the 18th Century* (Berkeley, CA: University of California Press, 1990).
2. E. H. Kossmann, *The Low Countries, 1780–1940* (Oxford: Clarendon Press, 1978), p. 124.

3. Kossmann, *The Low Countries*, p. 125.
4. M. Weber, in G. Roth and C. Wittich (eds), *Economy and Society: An Outline of Interpretive Sociology* (Berkeley, CA: University of California Press, 1978), p. 957.
5. R. André, 'Adolphe Quetelet, Académicien', *Actualité et universalité de la pensée scientifique d'Adolphe Quetelet: Actes du colloque organisé à l'occasion du bicentenaire de sa naissance* (Brussels: Académie royale de Belgique, 1997), pp. 23–45, on p. 23.
6. On the rise (and fall) of the Paris Observatory in particular, see D. Aubin, 'The Fading Star of the Paris Observatory in the Nineteenth Century: Astronomers' Urban Culture of Circulation and Observation', *Osiris*, 18 (2003), pp. 70–100. For an examination of the paradigmatic observatory of the era, see R. W. Smith, 'A National Observatory Transformed: Greenwich in the Nineteenth Century', *Journal for the History of Astronomy*, 22 (1991), pp. 5–29. On the public importance of observatories across Europe, see nearly every article in the exceptional C. Bigg, D. Aubin and H. O. Sibum (eds), *The Heavens on Earth: Observatories and Astronomy in Nineteenth-Century Science and Culture* (Durham, NC: Duke University Press, 2010), in particular T. Levitt, '"I Thought this Might Be of Interest...": The Observatory of Public Enterprise', pp. 285–304; O. Molvig, 'The Berlin Urania, Humboldtian Cosmology, and the Public', pp. 326–44; C. Bigg, 'Staging the Heavens: Astrophysics and Popular Astronomy in the Late Nineteenth Century', pp. 305–24; and the introduction by C. Bigg, D. Aubin and H. O. Sibum, pp. 1–32.
7. A. Collard, 'Adolphe Quetelet et l'astronomie', *Ciel et terre*, 6 (1935), pp. 1–10, on p. 1.
8. G. W. T. Ormond, 'The Question of the Netherlands, 1829–1830', *Transactions of the Royal Historical Society*, 2 (1919), pp. 150–71.
9. Ormond, 'The Question of the Netherlands, 1829–1830', p. 152.
10. A. Quetelet, 'Extrait d'un rapport sur la formation d'un observatoire dans le royaume des Pays-Bas', repr. in *Correspondance mathématique et physique* (*CMP*), 11 vols (1823;1825), Rare Books Collection, Boston Public Library, vol. 1, pp. 67–70, on p. 67.
11. Quetelet experienced his own brush with censorship when his *Astronomie populaire* was placed on the Catholic Church's *Index*. F. H. Hankins, *Adolphe Quetelet as Statistician* (New York: Columbia University Press, 1908), pp. 1–136, on p. 30.
12. Aside from sharing a birthplace, Laensberg shared many qualities with his later champion, including a dedication to the compilation of data. Laensberg, anticipating Quetelet's volumes on the position of various shooting stars, composed at his death a table of *all* of the available observations of stars from Ptolemy to Kepler.
13. Quetelet, 'Extrait d'un rapport', *CMP*, vol. 1, p. 68.
14. Quetelet, 'Extrait d'un rapport', *CMP*, vol. 1, p. 68.
15. The importance of order and stability inside the observatory has been noted as a 'token of stability, integrity, order [and] permanence'. J. A. Bennett, 'The English Quadrant in Europe: Instruments and the Growth of Consensus in Practical Astronomy', *Journal for the History of Astronomy*, 23 (1992), pp. 1–14, on p. 2.
16. Quetelet, 'Extrait d'un rapport', *CMP*, vol. 1, p. 68.
17. Porter has claimed the Belgian Revolution provided a 'new turn of thinking' towards order and stability, but Quetelet's remarks in 1823 appear to indicate that the idea of stability in the political order, while certainly reinforced by the revolution, had its roots in earlier political disruptions in the Belgium. T. M. Porter, 'The Mathematics of Society: Variation and Error in Quetelet's Statistics', *British Journal for the History of Science*, 18 (1985), pp. 51–69, on p. 58.
18. Quetelet, 'Extrait d'un rapport', *CMP*, vol. 1, p. 68.

19. E. H. Kossmann, *The Low Countries, 1780–1940* (Oxford: Clarendon Press, 1978), p. 67.
20. A. Collard, 'Le Centenaire de la création de l'Observatoire royal de Bruxelles', *Ciel et terre*, 42 (1926), pp. 209–23, on p. 210.
21. A. Quetelet, *Notice sur Antoine Reinhard Falck* (Brussels: Hayez, 1844).
22. Quetelet, *Notice sur Antoine Reinhard Falck*, p. 17.
23. Quetelet, 'Extrait d'un rapport', *CMP*, vol. 1, p. 69.
24. Quetelet, *Notice sur Antoine Reinhard Falck*, p. 30.
25. Quetelet, 'Extrait d'un rapport', *CMP*, vol. 1, p. 69.
26. Collard, 'Le Centenaire', p. 215. Collard's history reproduces most of Quetelet's correspondence and, with some exceptions, will be the primary reference for this section.
27. J. A. Bennett, *The Divided Circle: A History of Instruments for Astronomy, Navigation and Surveying* (Oxford: Phaidon-Christie's, 1987) and R. W. Smith, 'Remaking Astronomy: Instruments and Practice in the Nineteenth and Twentieth Centuries', in M. J. Nye (ed.), *Cambridge History of Science*, 7 vols, vol. 5: *The Modern Physical and Mathematical Science* (Cambridge: Cambridge University Press, 2002), pp. 154–73.
28. Bennett, 'The English Quadrant in Europe', p. 2. For a survey of the larger problem of superfluous data in positional astronomy, see K. Donnelly, 'On the Boredom of Science: Positional Astronomy in the Nineteenth Century', *British Journal for the History of Science*, 47 (2014), pp. 473–503, on pp. 493–7.
29. Collard, 'Le Centenaire', p. 216.
30. J. Bourgeois, 'Planet', in J. L. Heilbron (ed.), *The Oxford Guide to the History of Physics and Astronomy* (Oxford: Oxford University Press, 2005), pp. 253–6, on p. 254.
31. M. Daumas, *Scientific Instruments of the Seventeenth and Eighteenth Century*, trans. Mary Holbrook (New York: Praeger, 1972), p. 180.
32. [Anon.], 'Fraunhofer – Refractive and Dispersive Powers of Glass, and the Achromatic Telescope', *Foreign Quarterly Review*, 1 (1827), pp. 424–35, on p. 433.
33. A. Quetelet, 'Extrait d'un lettre contenant les observations de la comète découverte en novembre 1823, faite par MM. Bouvard et Nicolett', *CMP*, vol. 1, pp. 143–4, on p. 144. According to an obituary for one of the principal figures in developing the achromatic lens for the larger telescope, £950 was the price for the nine-inch diameter lens. [Anon.], 'Memoir of the Life of Joseph Fraunhofer', *Journal of the Franklin Institute*, 4 (1829), pp. 96–103, on p. 101. This would be the equivalent of roughly $150,000 today.
34. Collard, 'Le Centenaire', p. 217.
35. Collard, 'Le Centenaire', p. 221.
36. Collard, 'Le Centenaire', p. 220.
37. Kossmann, *The Low Countries*, pp. 115–25.
38. Collard, 'Adolphe Quetelet et l'astronomie', p. 7.
39. Collard, 'Le Centenaire', pp. 221.
40. A. Quetelet, *Des lois concernant le développement de l'homme* (Brussels: Hayez, 1870), p. 3.
41. Collard, 'Le Centenaire', p. 218.
42. For the various projects for which Quetelet would make use of the observatory, see below in the Conclusion.
43. E. H. Kossmann, *The Low Countries, 1780–1940* (Oxford: Clarendon Press, 1978), p. 123.
44. This note comes from the *avant-propos* of the third edition of *Astronomie élémentaire*, published nine years after the first edition of the lectures. A. Quetelet, *Astronomie élémentaire*, 3rd edn, 2 vols (1826; Brussels: Tircher, 1834), vol. 1, p. iii.
45. Quetelet, *Astronomie élémentaire*, vol. 1, p. iii.
46. Quetelet, *Astronomie élémentaire*, vol. 1, p. 26.

47. Quetelet, *Astronomie élémentaire*, vol. 1, p. 56.
48. Quetelet, *Astronomie élémentaire*, vol. 1, pp. 85, 76.
49. Quetelet, *Astronomie élémentaire*, vol. 1, p. 88.
50. Quetelet, *Astronomie élémentaire*, vol. 1, p. 90.
51. Ore has even gone so far as to suggest that Cardano's other interests kept him from being identified as the true founder of probability theory, a full century prior to Pascal. O. Ore, *Cardano the Gambling Scholar* (Princeton, NJ: Princeton University Press, 1953).
52. Conversely, it has been argued that it was the desire to improve telescopes themselves that led to more ambitious empirical questions. R. Smith, 'Remaking Astronomy: Instruments and Practice in the Nineteenth and Twentieth Centuries', in M. J. Nye (ed.), *Cambridge History of Science*, 7 vols, vol. 5: *The Modern Physical and Mathematical Science* (Cambridge: Cambridge University Press, 2002), pp. 154–73, on p. 156.
53. A. Quetelet, *Astronomie populaire*, 2nd edn (1827; Brussels: Remy, 1832), p. vi.
54. O. Molvig, 'The Berlin Urania, Humboldtian Cosmology, and the Public', in C. Bigg, D. Aubin and H. O. Sibum (eds), *The Heavens on Earth: Observatories and Astronomy in Nineteenth-Century Science and Culture* (Durham, NC: Duke University Press, 2010), pp. 325–44, on pp. 336, 333.
55. Quetelet, *Astronomie populaire*, p. vi.
56. Though at least one Belgian has argued that Quetelet's influence was so profound that 'one could say he created the Academy like he created the observatory'. M.A. Demoulin, 'Adolphe Quetelet, fondateur de l'Observatoire royal de Belgique', *Bulletin astronomique de l'Observatoire royal de Belgique*, 2 (1935), pp. 1–3, on p. 3.
57. J. Lavalleye, *L'Académie royale des sciences, des lettres et des beaux-Arts, 1772–1972: Esquisse Historique* (Brussels: Palais Des Académies, 1972), p. 51.
58. Lavalleye, *L'Académie royale*, p. 16.
59. Lavalleye, *L'Académie royale*, p. 51.
60. A. Quetelet, *Sciences mathématique et physique chez les belges, au commencement du XIXe siècle* (Brussels: Buggenhoudt, 1866), p. 728.
61. Lavalleye, *L'Académie royale*, p. 52.
62. Lavalleye, *L'Académie royale*, p. 48.
63. Lavalleye, *L'Académie royale*, p. 51.
64. Quetelet, *Sciences mathématique et physique*, p. 729.
65. Quetelet, *Sciences mathématique et physique*, p. 329.
66. [Anon.], 'Journal des séances', *Nouveaux mémoire de l'Académie royale des sciences et des belles-lettres de Bruxelles*, 3 (1826), pp. i–xxiv.
67. Quetelet, *Sciences mathématique et physique*, p. 246.
68. Lavalleye, *L'Académie royale*, p. 52.
69. Quetelet, *Sciences mathématique et physique*, pp. 315, 290.
70. Quetelet, *Sciences mathématique et physique*, p. 271.
71. Quetelet, *Sciences mathématique et physique*, p. 268.
72. Quetelet, *Sciences mathématique et physique*, p. 269.
73. G. Dewalque, 'François-Philippe Cauchy', in *Biographie nationale* (Brussels: Bruylant, 1872), vol. 3, pp. 380–3, on p. 380.
74. Quetelet, *Sciences mathématique et physique*, p. 253.
75. Quetelet, *Sciences mathématique et physique*, pp. 94, 185.
76. Quetelet, *Sciences mathématique et physique*, p. 281.
77. Quetelet, *Sciences mathématique et physique*, p. 280.
78. Quetelet, *Sciences mathématique et physique*, p. 259.

79. Lavalleye, *L'Académie royale*, p. 60.
80. Quoted in Lavalleye, *L'Académie royale*, p. 39.
81. A. Quetelet, 'Quelques remarques sur l'influence des académies, des congrès, et des conférences scientifique' (1853), *Quetelet Varia* (*QV*), Bibliothèque royale de Belgique, vol. 4, #9, p. 1.
82. Quetelet, 'Quelques remarques sur l'influence, p. 2.
83. Quetelet, 'Quelques remarques sur l'influence, p. 3.
84. Quetelet, 'Quelques remarques sur l'influence, p. 3.
85. Quoted in Lavalleye, *L'Académie royale*, p. 56.
86. Lavalleye, *L'Académie royale*, p. 56.
87. Quetelet, *Sciences mathématique et physique*, p. 271.
88. P. Bourdieu, 'The Peculiar History of Scientific Reason', *Sociological Forum*, 6 (1991), pp. 3–26, on p. 16.

4 From Brussels to Europe: The Creation of a Scientific Network, 1823–9

1. A. de Tocqueville, *The Old Regime and the Revolution*, trans. S. Gilbert (French, 1856; Chicago, IL: University of Chicago Press, 1998), p. 139.
2. R. Fox, *The Savant and the State: Science and Cultural Politics in Nineteenth-Century France* (Baltimore, MD: Johns Hopkins University Press, 2011), p. 2.
3. F. H. Hankins, *Adolphe Quetelet as Statistician* (New York: Columbia University Press, 1908), 1–136, on p. 21.
4. A. Quetelet, *Notice sur Alexis Bouvard* (Brussels: Hayez, 1844), *Quetelet Varia* (*QV*), Bibliothèque royale de Belgique, vol. 4, #118, p. 7.
5. Quetelet, *Notice sur Alexis Bouvard*, p. 7.
6. P. Pâquet, 'Les Initiatives de Adolphe Quetelet en astronomie', *Actualité et universalité de la pensée scientifique d'Adolphe Quetelet: Actes du colloque organisé à l'occasion du bicentenaire de sa naissance* (Brussels: Académie royale de Belgique, 1997), pp. 63–72.
7. A. Collard, 'Le Centenaire de la création de l'Observatoire royal de Bruxelles', *Ciel et terre*, 42 (1926), pp. 209–23, on p. 212.
8. Quetelet, *Notice sur Alexis Bouvard*, p, 8.
9. Stigler attributes much of it to the 'unusual force' of Quetelet's personality, which is no doubt true: 'so bright was the light in Quetelet's eyes, so strong the vision he conveyed, and so impressed with him were all who met him, that against all odds he was successful'. S. Stigler, 'Adolphe Quetelet: Statistician, Scientist, Builder of Intellectual Institutions', *Actualité et universalité de la pensée scientifique d'Adolphe Quetelet: Actes du colloque organisé à l'occasion du bicentenaire de sa naissance* (Brussels: Académie royale de Belgique, 1997), pp. 47–61, p. 49.
10. M. Bakich, *The Cambridge Planetary Handbook* (Cambridge: Cambridge University Press, 2000), p. 279.
11. Quetelet, *Notice sur Alexis Bouvard*, p. 7.
12. A. Quetelet, 'Dominique-François-Jean Arago', *Annuaire de l'Académie royale de Belgique* (1855), *QV*, vol. 4, #127/128, p. 6.
13. Quetelet, *Notice sur Alexis Bouvard*, p. 9.
14. This was due likely to Arago's unceasing fight against sinecures, careerism and nepotism.
15. M. Daumas, *Arago, 1786–1853: La Jeunesse de la science*, 2nd edn (Paris: Belin, 1987), p. 33.

16. Daumas, *Arago, 1786–1853*, p. 33.
17. D. Outram, 'Politics and Vocation: French Science, 1793–1830', *British Journal for the History of Science*, 12 (1980), pp. 27–43.
18. See above, Chapter One.
19. Quetelet, 'Dominique-François-Jean Arago', p. 25.
20. Quetelet, 'Dominique-François-Jean Arago', p. 10.
21. On the *cumul*, see I. Grattan-Guiness, *Convolutions in French Mathematics: From the Calculus and Mechanics to Mathematical Analysis and Mathematical Physics*, 3 vols (Basel: Birkhäuser, 1990), vol. 2, pp. 1271–3 and M. Crosland, *Scientific Institutions and Practices in France and Britain, c.1700–c.1850* (Burlington, VT: Ashgate, 2007), p. 276.
22. D'Alembert claimed that 'We were tested at the Académie des sciences, myself and M. Condorcet, by the harassments which disgusted us'. Quoted by Quetelet, 'Dominique-François-Jean Arago', p. 12n.
23. M. Crosland, *Science under Control: The French Academy of Sciences, 1795–1914* (1992; Cambridge: Cambridge University Press, 2002), p. 372.
24. Quetelet, 'Dominique-François-Jean Arago', p. 14.
25. J. Dhombres, 'Preface', in Daumas, *Arago, 1786–1853*, pp. 5–9, on p. 5.
26. Daumas, *Arago, 1786–1853*, p. 141.
27. Quetelet, 'Dominique-François-Jean Arago', p. 35. On the importance of the relatively new idea of 'disinterest' in science, see L. Daston and P. Galison, *Objectivity* (New York: Zone, 2007), pp. 115–91.
28. Quetelet, 'Dominique-François-Jean Arago', p. 36.
29. Quetelet, 'Dominique-François-Jean Arago', p. 36.
30. Quetelet, 'Dominique-François-Jean Arago', p. 37.
31. *Correspondance mathématique physique* (*CMP*), 11 vols (1825; Brussels: Hayez, 1825), vol. 1, p. 1.
32. *CMP*, vol. 1, p. 2.
33. *CMP*, vol. 1, p. 144.
34. *CMP*, vol. 1, pp. 101, 217, 116, 231.
35. For the importance of the relationship between Villermé and Quetelet for both men's careers, see K. Donnelly, 'Social Physics or Social Disease: Villermé, Quetelet and Cholera in Paris and Brussels, 1832', in E. Martone (ed.), *Royalists, Radicals, and les Misérables: France in 1832* (Newcastle: Cambridge Scholars Press, 2013), pp. 99–119.
36. *CMP*, vol. 2 (1826), p. 286.
37. *CMP*, vol. 3 (1827), p. 263.
38. Though many European thinkers at the time proposed the introduction of scientific methodology into social research, most plans to expand the methodology of the natural science were – owing to the influence of the *Ideologues* and positivists – decidedly anti-quantitative.
39. A. Quetelet, *Recherches sur la population, les naissances, les décès, les dépôts de mendicité, etc. dans le Royaume des Pays-Bas* (Brussels: Tarlier, 1827), p. 1.
40. Quetelet, *Recherches sur la population*, p. 2.
41. For the European production of the 'avalanche of numbers', see I. Hacking, *The Taming of Chance* (Cambridge: Cambridge Press, 1990), pp. 16–46.
42. *CMP*, vol. 1, p. 18.
43. *CMP*, vol. 1, p. 16.
44. Baily in fact would later join with a number of 'business astronomers', including several Quetelet knew like John Herschel and Charles Babbage to lobby for the precision of

astronomy in all walks of life. It is possible that there was a significant overlap in interests, especially since the business astronomers looked to 'the French analytical system of institutional administration ... [for] calculation and analysis as a set of value-free technocratic techniques'. W. J. Ashworth, 'The Calculating Eye: Baily, Herschel, Babbage and the Business of Astronomy', *British Journal for the History of Science*, 27 (1994), pp. 409–41, on p. 414. In spite of personal connections with a host of British men of science, and his participation in the creation of the Statistical Society for the British Association for the Advancement of Science, Quetelet had a very different vision of science and statistics from his Victorian counterparts. In particular, he rarely shared their concern with business, industry or, it seems, money. On the association of Quetelet and the British statistical group known as the Cambridge Network, see S. F. Cannon, *Science in Culture: The Early Victorian Period* (New York: Science History Publications, 1978), pp. 241–5. However, it has been argued that 'in the course of the 1830s the founding group undoubtedly came to consider the Statistical Society of London a failure'. L. Goldman, 'The Origins of British "Social Science": Political Economy, Natural Science and Statistics, 1830–1835', *Historical Journal*, 26 (1983), pp. 587–616, on p. 608.

45. F. Loise, 'Jean-François Lemaire', *Biographie nationale* (Brussels: Bruylant, 1891), vol. 11, pp. 779–81, on p. 779.
46. *CMP*, vol. 2, p. 230.
47. *CMP*, vol. 2, p. 126.
48. *CMP*, vol. 2, p. 287.
49. *CMP*, vol. 2, p. 232.
50. *CMP*, vol. 2, p. 232.
51. *CMP*, vol. 2, p. 286.
52. See above, Chapter Three.
53. *CMP*, vol. 2, p. 249.
54. *CMP*, vol. 3, p. 85.
55. *CMP*, vol. 3, p. 236.
56. *CMP*, vol. 5 (1829).
57. *CMP*, vol. 5.
58. *CMP*, vol. 1, p. 18.
59. A. Quetelet, *Notes extraits d'un voyage scientifique fait en Allemagne pendant l'été de 1829* (Brussels: Hayes, 1830).
60. A. Quetelet, *Sciences mathématique et physique chez les belges au XIXe siècle* (Brussels: Buggenhoudt, 1866), p. 630.
61. Quetelet, *Notes extraits*, p. 7.
62. Quetelet, *Sciences mathématique et physique*, p. 637.
63. Quetelet, *Sciences mathématique et physique*, p. 637.
64. Quetelet, *Sciences mathématique et physique*, p. 635.
65. Quetelet, *Notes extraits*, p. 13.
66. Quetelet, *Sciences mathématique et physique*, p. 14.
67. Quetelet, *Sciences mathématique et physique*, p. 23.
68. Quetelet, *Sciences mathématique et physique*, p. 22.
69. The composer was only twenty at the time, and Quetelet noted him simply as the 'son of the great philosopher', Moses Mendelssohn.
70. Quetelet, *Sciences mathématique et physique*, p. 592.
71. Quetelet, *Sciences mathématique et physique*, p. 592.
72. K. Donnelly, 'On the Boredom of Science: Positional Astronomy in the Nineteenth

Century', *British Journal for the History of Science*, 47 (2014), pp. 473–503, on p. 474.
73. Quetelet, *Sciences mathématique et physique*, p. 604.
74. Quetelet, *Notes extraits*, p. 25.
75. Quetelet, *Sciences mathématique et physique*, p. 661.
76. D. Sepper, *Goethe contra Newton: Polemics and the Project for a New Science of Color* (Cambridge: Cambridge University Press, 1988), p. 2.
77. Sepper calls Goethe's method 'rigorous and empirical', yet one that searched 'for unities'. D. Sepper, 'Goethe, Colour and the Science of Seeing', in A. Cunningham and N. Jardine (eds), *Romanticism and the Sciences* (Cambridge: Cambridge University Press, 1990), pp. 189–98, on p. 192.
78. Quetelet, *Sciences mathématique et physique*, p. 662.
79. For an account of this exchange, see A. Collard, 'Goethe et Quetelet: Leurs relations de 1829 à 1832', *Isis*, 20 (1934), pp. 426–35.
80. G. W. Dunnington, *Carl Friedrich Gauss: Titan of Science* (New York: Mathematical Association of America, 1955), p. 242.
81. Quetelet, *Sciences mathématique et physique*, p. 645.
82. Gauss to Olbers, 12 October 1829. Quoted in G. W. Dunnington, *Carl Friedrich Gauss: Titan of Science* (New York: Mathematical Association of America, 1955), p. 154.
83. Gauss's interest in Quetelet may not be so surprising. Gauss's earlier biographer claimed that 'his derivation of the error curve ... probably started from his own empirical results in measuring'. T. Hall, *Carl Friedrich Gauss*, trans. Albert Froderberg (Cambridge, MA: Massachusetts Institute of Technology Press, 1970), p. 78. This is supported in T. M. Porter, *Trust in Numbers* (Princeton, NJ: Princeton University Press, 1995).
84. Quetelet, *Sciences mathématique et physique*, p. 650.
85. Terrestrial magnetism would be one of many earth sciences where Quetelet's Brussels Observatory took a leading role. See Conclusion.
86. Quetelet, *Notes extraits*, p. 51.
87. Quetelet, *Sciences mathématique et physique*, p. 663.
88. See above, Chapter Three.
89. A. Quetelet, 'Quelques remarques sur l'influence des académies, des congrès, et des conférences scientifique' (1853), *Quetelet Varia* (*QV*), Bibliothèque royale de Belgique, vol. 4, #9, p. 5.
90. Quetelet, 'Quelques remarques', p. 6.

5 *Physique Sociale*, 1825–35

1. C.-A. Helvétius, *De l'esprit* (Paris: Durand, 1758), p. 6. Translation quoted from Condorcet, 'A General View of the Sciences of Social Mathematics', in *Condorcet: Selected Writings*, ed. and trans. K. M. Baker (French, 1793; Indianapolis, IN: Bobbs-Merrill, 1976), pp. 183–206, on p. 190.
2. Much to the dismay of Auguste Comte, who accused Quetelet of stealing the term *physique sociale*, forcing the Frenchman to adopt his less preferred term *sociologie*. As Comte wrote: 'This term, no less indispensable than 'positive philosophy', was first used seventeen years ago in my first works on political philosophy. However recently these two essential terms have been in some sense spoiled by the vicious attempts of different writers to appropriate them ... I must above all signal the abuse with regard to the first term, which was adopted by a *savant belge* in recent years as a title of a work which is about nothing more than simple statistics.' A. Comte, *Cours de philosophie positive*, 6

vols (Paris: Bachelier, 1839), vol. 4, p. 7, n. 1.
3. K. M. Baker, *Condorcet: From Natural Philosophy to Social Mathematics* (Chicago, IL: University of Chicago Press, 1975), p. 372. On Laplace, see T. M. Porter, 'The Mathematics of Society: Variation and Error in Quetelet's Statistics', *British Journal for the History of Science*, 18 (1985), pp. 51–69, on p. 51. Some have even mistakenly referred to Quetelet as 'Laplace's pupil'. S. Lalvani, *Photography, Vision, and the Production of Modern Bodies* (Albany, NY: State University of New York Press, 1996), p. 100. It has also been pointed out that Hankins's original mention of the relationship between Laplace and Quetelet included no source. S. Stigler, *The History of Statistics: The Measurement of Uncertainty before 1900* (Cambridge, MA: Harvard University Press, 1986), p. 162, n. 4.
4. Exceptions to this included both the 1869 edition of *Physique sociale* and *Anthropométrie* (1871), books that consisted mostly of repackaged papers from earlier in Quetelet's career.
5. While Quetelet's claim that social physics was only a 'distraction' from astronomy was probably an overstatement, he never again focused exclusively on social statistics. See Chapter Three.
6. Porter, 'The Mathematics of Society', p. 58.
7. The earliest English reference for this moniker dates to at least 1875, when the head of the United States Bureau of Statistics referred to Quetelet as a notable exception to the 'regrettable' failures of nineteenth-century statistical organizations. E. Young, *Labor in Europe and America* (Philadelphia, PA: S.A. George and Co., 1875), p. 645. Young, however, uses the name in quotation marks to suggest that it had at least some general currency. In French, the reference to Quetelet as 'le père de la statistique moderne' appears in a search of Google Books as early as 1908. E. Francken and E. Mahaim, 'La Loi d'erreur de M. F. Y. Edgeworth', *Revue universelle des mines, de la métallurgie des travaux publics, des sciences et des arts appliqués à l'industrie*, 22 (1908), pp. 219–66, on p. 227. While most scholarship has abandoned attempts to assign paternity to ideas or disciplines, Quetelet's title has endured. For a recent example, see G. Gill, *We Two: Victoria and Albert: Rulers, Partners, Rivals* (New York: Random House, 2009), p. 134.
8. J. Graunt, 'Natural and Political Observations upon the Bills of Mortality (1665)', in C. H. Hull (ed.), *The Economic Writings of Sir William Petty* (Ithaca, NY: Cornell University Press, 1963), pp. 314–435, on p. 333.
9. Stigler's work is again an exception to this rule.
10. For Quetelet's role in the fight see I. B. Cohen, 'Revolutions, Scientific, and Probabilistic', in L. Krüger, L. J. Daston and M. Heidelberger (eds), *The Probabilistic Revolution* (Cambridge, MA: Massachusetts Institute of Technology Press, 1987), pp. 23–44, on pp. 37–40 and A. Oberschall, 'The Two Empirical Roots of Social Theory and the Probability Revolution', in G. Gigerenzer, L. Krüger and M. S. Morgan (eds), *The Probabilistic Revolution*, 2 vols, vol 2: *Ideas in the Sciences* (Cambridge, MA: Massahusetts Institute of Technology Press, 1987), pp. 103–34, on pp. 108–12.
11. I. Todhunter, *A History of the Mathematical Theory of Probability from the Time of Pascal to That of Laplace* (Cambridge: Macmillan & Co., 1865).
12. I. Hacking, *The Emergence of Probability: A Philosophical Study of Early Ideas and Probability, Induction and Statistical Interference* (Cambridge: Cambridge University Press, 1984) and S. Stigler, *The History of Statistics: The Measurement of Uncertainty before 1900* (Cambridge, MA: Harvard University Press, 1986).
13. T. M. Porter, *The Rise of Statistical Thinking, 1820–1900* (Princeton, NJ: Princeton University Press, 1986).
14. For the technical, as opposed to social and political, developments in probability, see

chapter 4 of Stigler, *The History of Statistics*, pp. 139–58.
15. L. Daston, *Classical Probability in the Enlightenment* (Princeton, NJ: Princeton University Press, 1988), pp. 48–111.
16. Todhunter, *A History of the Mathematical Theory of Probability*, pp. 11–14.
17. To imagine the difference is to picture a scenario in which it was known that an American had a .001 per cent chance of drinking fatally poisoned tap water. This information would be negligible to an individual deciding whether to drink, but in a population of 300 million, this would mean that 3,000 Americans would die. The information is therefore crucial for the state but of little use for 'the reasonable man'.
18. M. Raeff, *The Well-Ordered Police State: Social and Institutional Change through Law in the Germanies and Russia, 1600–1800* (New Haven, CT: Yale University Press, 1983), J. C. Scott, *Seeing like a State: How Certain Schemes to Improve the Human Condition Have Failed* (New Haven, CT: Yale University Press, 1998) and J. Caplan and J. Torpey (eds), *Documenting Individual Identity: The Development of State Practices in the Modern World* (Princeton, NJ: Princeton University Press, 2001).
19. On attempts at qualitative and quantitative statistics near Quetelet, see P. M. M. Klep and I. H. Stamhuis, 'Introduction', in P. M. M. Klep and I. H. Stamhuis (eds), *The Statistical Mind in a Pre-Statistical Era: The Netherlands, 1750–1850* (Amsterdam: Askant, 2002), pp. 13–28, on p. 17.
20. Stigler, *The History of Statistics*, pp. 5–7.
21. Daston, *Classical Probability in the Enlightenment*, p. 372.
22. A. Quetelet, *Instructions populaires sur le calcul des probabilités* (Brussels: Tarlier and Hayez, 1828).
23. A. Quetelet, 'Mémoire sur les lois des naissances et de la mortalité à Bruxelles', *Nouveaux mémoires de l'Académie*, 3 (1826), pp. 493–513, on p. 495.
24. Quetelet, 'Mémoire sur les lois', p. 496n.
25. Quetelet, 'Mémoire sur les lois', pp. 497–8.
26. Quetelet's data excluded 1815 and the 'terrible catastrophe' of Waterloo.
27. In actuality, Quetelet had to alter his data slightly to achieve this relationship. While the gross number of births peaked in February, death rates were only highest in February on a 'per-day' basis.
28. T. R. Malthus, *Malthus: An Essay on the Principle of Population*, ed. D. Winch (1798; Cambridge: Cambridge University Press, 1992), pp. 13–14.
29. Quetelet, 'Mémoire sur les lois', p. 500.
30. For one of the most interesting readings of Quetelet and Malthus, which sees Quetelet as the means by which Charles Darwin encountered Malthusian ideas, see S. S. Schweber, 'The Origin of the *Origin* Revisited', *Journal for the History of Biology*, 10 (1977), pp. 229–316. For a contrary view, see A. Ariew, 'Under the Influence of Malthus's Law of Population Growth: Darwin Eschews the Statistical Techniques of Adolphe Quetelet', *Study of History and Philosophy of Biological and Biomedical Sciences*, 38 (2007), pp. 1–19.
31. Quetelet, 'Mémoire sur les lois', p. 506.
32. Quetelet, 'Mémoire sur les lois', p. 496.
33. *Correspondance mathématique physique* (*CMP*), 11 vols (1825; Brussels: Hayez, 1825), vol. 1, p. 16
34. See above, Chapter Four.
35. A. Quetelet, *Recherches sur la population, les naissances, les décès, les dépôts de mendicité, etc. dans le Royaume des Pays-Bas* (Brussels: Tarlier, 1827), p. 8.
36. Quetelet, *Recherches sur la population*, p. 13.

37. Quetelet, *Recherches sur la population*, p. 13.
38. Quetelet relied here on the Newtonian move of not positing ultimate causation. For a summary of the difference between a law *causing* something and a law *being* something in Newton, see P. Fara, *Newton: The Making of Genius* (New York: Columbia University Press, 2007), pp. 104–5. A longer philosophical analysis with guides to further references can be found in A. Janiak, 'Newton and the Reality of Force', *Journal of the History of Philosophy*, 45 (2007), pp 127–47.
39. Quetelet, *Recherches sur la population*, p. 10.
40. *CMP*, vol. 2, p. 286.
41. Quetelet, *Recherches sur la population*, p. 16.
42. Quetelet, *Recherches sur la population*, p. 17.
43. Quetelet, *Recherches sur la population*, p. 18.
44. Quetelet, *Recherches sur la population*, p. 33.
45. Quetelet, *Recherches sur la population*, p. 41.
46. Quetelet, *Recherches sur la population*, p. 47.
47. Quetelet, *Recherches sur la population*, p. 60.
48. Quetelet, *Recherches sur la population*, p. 66.
49. Quetelet, *Recherches sur la population*, p. 67.
50. Quetelet, *Recherches sur la population*, p. 55.
51. Quetelet, *Recherches sur la population*, p. 52.
52. A. Quetelet, *Sur l'homme et le développement de ses facultés*, 2 vols (Brussels: Bachelier, 1835), vol. 2, p. 165.
53. Quetelet, *Recherches sur la population*, pp. 67–8.
54. Though certainly not monolithic, the writers of eighteenth-century France provided a form of progressive history that was well suited to Quetelet's plans for social physics. For a summary of the many Enlightenment perspectives on history, see chapter 2 of M. Hulliung, *The Autocritique of Enlightenment: Rousseau and the Philosophes* (Cambridge, MA: Harvard University Press, 1994).
55. A. Quetelet, *Recherches statistique* (Brussels: Hayez, 1829), p. i.
56. Quetelet, *Recherches statistique*, p. iv.
57. Quetelet, *Recherches statistique*, p. ii.
58. Here Quetelet might be distinguished from earlier attempts of probabilist thinkers like Turgot, Laplace and Condorcet, whose application of statistics to society have been described as 'despotic'. C. C. Gillispie, 'Probability and Politics', *Proceedings of the American Philosophical Society*, 116 (1972), pp. 1–20, on p. 19.
59. Quetelet, *Recherches statistique*, p. iv.
60. L. W.-De Donder, 'Une lettre de Cécile Quetelet relative à la révolution belge de 1830', *Cahiers Bruxellois*, 6 (1961), pp. 68–75.
61. T. M. Porter, *The Rise of Statistical Thinking, 1820–1900* (Princeton, NJ: Princeton University Press, 1986), p. 58.
62. A. Quetelet, *Recherches sur le penchant au crime aux différens âges*, 2nd edn (1831; Brussels: Hayez, 1833), p. 2.
63. A. Quetelet, *Recherches sur la loi de la croissance de l'homme* (Brussels: Hayez, 1831), p. 13.
64. Quetelet, *Recherches sur la loi de la croissance de l'homme*, pp. 20–5.
65. A. Quetelet, *Physique sociale ou essai sur le développement des facultés de l'homme*, ed. É. Vilquin and J.-P. Sanderson, new edn (1869; Brussels: Académie royale de Belgique, 1995), p. 324 (1–28). Vilquin and Sanderson have benefitted scholars immensely by keeping the original pagination in their annotated edition. The number in parentheses

refers to the 1869 edition volume and page.
66. Statistically speaking, Quetelet's association between the two distributions was mistaken. Gauss's curve was meant to show observational error in the determination of a fixed star point, while the 'error' found in height and weight was only in relation to an abstract ideal. There really was a star; 'true' height or weight was only an artefact of the data.
67. In 1972, Ancel Keys found that the relationship Quetelet drew between height and weight was the best large-scale predictor of obesity. He renamed it the body mass index. F. Fidanza et al., 'Indices of Relative Weight and Obesity', *Journal of Chronic Diseases*, 25 (1972), pp. 329–43.
68. A. Quetelet, *Recherches sur le poids de l'homme aux différens âges* (Brussels: Hayez, 1833), p. 27.
69. Quetelet, *Recherches sur le poids de l'homme aux différens âge*, p. 28.
70. While great debates continue about the relevance of the BMI for individuals, studies of obesity continue to reference Quetelet's formula, believing it to be the most accurate means to determine health risks for large populations. In 2007, one commenter noted that the BMI 'remains a dependable value and the basis of much of the associations reported heretofore with obesity'. G. Eknoyan, 'Adolphe Quetelet (1796–1874) – the Average Man and Indices of Obesity', *Nephrol Dial Transplant*, 23 (2008), pp. 47–51. For confirmation of the BMI's success, see T. Khosla and C. R. Lowe, 'Indices of Obesity Derived from Body Weight and Height', *British Journal of Preventive and Social Medicine*, 21 (1967), pp. 122–8. On problems with the BMI, a good general overview can be found in P. Bee, 'The Bmi Myth', *The Guardian* (27 November 2006), p. 18.
71. A. Quetelet, *Recherches sur le penchant au crime aux différens âges*, 2nd edn (1831; Brussels: Hayez, 1833), p. 4.
72. A. Quetelet, *Physique sociale ou essai sur le développement des facultés de l'homme*, ed. É. Vilquin and J.-P. Sanderson, new edn (1869; Brussels: Académie royale de Belgique, 1995), p. 419 (2–154).
73. See above, Chapter Three.
74. Quetelet, *Recherches sur le penchant au crime*, p. 5.
75. Intelligence and courage are measured quantitatively today, the former most obviously in IQ tests. On quantitative measures of courage, see P. J. Norton and B. J. Weiss, 'The Role of Courage on Behavioral Approach in a Fear-Eliciting Situation: A Proof-of-Concept Pilot Study', *Journal of Anxiety Disorders*, 23 (2009), pp. 212–17. There is by no means agreement on the value (or morality) of such research. For the most comprehensive and well-known history that attempts to debunk such efforts, see S. J. Gould, *The Mismeasure of Man (Revised and Expanded)* (1981; New York: W. W. Norton & Co., 2008). Gould does not mention Quetelet in his survey, starting only with Francis Galton as the 'apostle of quantification'.
76. While the 1842 English version of *Sur l'homme* translated 'exagéré' as 'absurd', Quetelet's 1831 paper had used the stronger 'absurdité' in his hypothetical objection to *Physique sociale*.
77. Quetelet, *Physique sociale*, p. 413 (2–144).
78. Quetelet, *Physique sociale*, p. 413 (2–145).
79. Quetelet, *Physique sociale*, p. 413 (2–145).
80. Quetelet, *Physique sociale*, p. 417 (2–151).
81. A. Quetelet, *Sur l'homme et le développement de ses facultés*, 2 vols (Brussels: Bachelier, 1835), vol. 2, pp. 279–80.

6 The Other Average Man: *L'Homme Moyen* and its Critics

1. F. Dostoevsky, *Crime and Punishment*, trans. R. Pevear and L. Volokhonksy (Russian, 1867; London: Vintage, 2007), p. 50.
2. A Russian translation of *Sur l'homme* appeared in 1865 and received significant press around the time Dostoevsky wrote *Crime and Pubishment*. Dostoevsky, *Crime and Punishment*, p. 555, n. 33. On Quetelet and Dostoevsky, see I. Paperno, *Suicide as a Cultural Institution in Dostoevsky's Russia* (Ithaca, NY: Cornell University Press, 1997), pp. 23–5. Dostoevsky however reserved most of his anger for Henry Thomas Buckle. See I. Hacking, *The Taming of Chance* (Cambridge: Cambridge University Press, 1990), p. 125–6 and below.
3. There are many meanings for 'determinism' in the various European languages. In what follows the term is used in the sense made most famous by Laplace's demon, as the possibility that all future actions may be known through the investigation of past action in combination with the knowledge of laws. For a good survey of the meanings of the term, see I. Hacking, 'Nineteenth Century Cracks in the Concept of Determinism', *Journal of the History of Ideas*, 44 (1983), pp. 455–75, on pp. 457–64.
4. So strong is this legacy that Quetelet's work is often glossed as a Laplacian clone, as 'a follower of Laplace who was interested in deterministic laws', or that his 'justification for the search for statistical laws ... reflect his mentor (Laplace's) views'. S. S. Schweber, 'The Origin of the *Origin* Revisited', *Journal for the History of Biology*, 10 (1977), pp. 229–316, on p. 287.
5. [Anon.], 'Review of *Recherches sur le penchant au crime aux différens âges*', *Bulletin des sciences géographiques, économique politique, voyages*, 28 (1831), pp. 113–36, on p. 114.
6. [Anon.], 'Review', p. 118.
7. A. Quetelet, *Physique sociale ou essai sur le développement des facultés de l'homme*, ed. É. Vilquin and J.-P. Sanderson, new edn (1869; Brussels: Académie royale de Belgique, 1995), p. 570 (2–370).
8. Quetelet, *Physique sociale*, p. 570 (2–371).
9. Quetelet, *Physique sociale*, p. 570 (2–371).
10. Quetelet admitted that his distinctions were arbitrary and that another person could rank them another way.
11. Quetelet, *Physique sociale*, p. 438 (2–180).
12. Quetelet, *Physique sociale*, p. 440 (2–183).
13. Quetelet, *Physique sociale*, p. 442 (2–185, 186).
14. Quetelet, *Physique sociale*, p. 444 (2–189).
15. Quetelet, *Physique sociale*, p. 444 (2–189).
16. One of Quetelet's many legacies, unexplored here but of course of great relevance to Dostoevsky, was that 'Quetelet ... opened up the possibility of a sociological analysis of crime'. For an especially thorough summary, see P. Beirne, 'Adolphe Quetelet and the Origins of Positivist Criminology', *American Journal of Sociology*, 92 (1987), pp. 1140–69, on p. 1166.
17. Quetelet, *Physique sociale*, p. 474 (2–233).
18. Quetelet, *Physique sociale*, p. 476 (2–235).
19. Quetelet, *Physique sociale*, p. 476 (2–236).
20. Quetelet, *Physique sociale*, p. 476 (2–236).
21. Quetelet, *Physique sociale*, p. 484 (2–246).
22. Quetelet, *Physique sociale*, p. 485 (2–287).

23. For two accounts of what determinism meant at the time, see E. Cassirer, *Determinism and Indeterminism in Modern Physics: Historical and Systematic Studies of the Problem* (New Haven, CT: Yale University Press, 1956) and I. Hacking, 'Nineteenth Century Cracks in the Concept of Determinism', *Journal of the History of Ideas*, 44 (1983), pp. 455–75.
24. As one historian rightly noted, Quetelet was 'hopelessly ambiguous on the question of freedom, contingency, and determinism'. L. Krüger, 'The Slow Rise of Probabilism', in L. Krüger, L. J. Daston and M. Heidelberger (eds), *The Probabilistic Revolution*, 2 vols, vol. 1: *Ideas in History* (Cambridge, MA: Massachusetts Institute of Technology Press, 1987), pp. 59–89, on p. 67.
25. A. Quetelet, *Sur l'homme et le développement de ses facultés*, 2 vols (Brussels: Bachelier, 1835), vol. 1, p. 111.
26. Although Lottin provides a well-documented case for Quetelet's vision of free will following the 'autodeterminisation' of the eclectic philosophers, the strength of the case is due far more to Lottin's clever reading of Quetelet than the Belgian's acknowledgement. Quetelet's interest in eclecticism was much more limited than Lottin suggests. For Lottin's theory, see J. Lottin, *Quetelet, statisticien et sociologue* (Louvain: Institut supérieur de philosophie, 1912), p. 447.
27. The most cited work on *Queteletismus* comes from I. Hacking, *The Taming of Chance* (Cambridge: Cambridge University Press, 1990), pp. 125–31 and T. M. Porter, *The Rise of Statistical Thinking 1820–1900* (Princeton, NJ: Princeton University Press, 1986), pp. 151–92. See also, M. Bulmer, *Francis Galton: Pioneer of Heredity and Biometry* (Baltimore, MD: Johns Hopkins University Press, 2003), p. 173 and S. Kern, *A Cultural History of Causality: Science, Murder Novels, and Systems of Thought* (Princeton, NJ: Princeton University Press, 2004), p. 292.
28. In one history, a triumph in sociology comes when Queteletism is 'defeated'. S. P. Turner, *The Search for a Methodology of Social Science* (Boston, MA: D. Reidel Pub. Co., 1986), p. 99. For a more recent popular work that adopts this position uncritically, see E. J. Oliver, *Fat Politics: The Real Story Behind America's Obesity Epidemic* (Oxford: Oxford University Press, 2006), p, 17.
29. B.-P. Lécuyer, 'Probability in Vital and Social Statistics: Quetelet, Farr, and the Bertillons', in L. Krüger, L. J. Daston and M. Heidelberger (eds), *The Probabilistic Revolution*, 2 vols, vol. 1: *Ideas in History* (Cambridge, MA: Massachusetts Institute of Technology Press, 1987), pp. 317–35, on p. 318.
30. For a nice summary of the project see H. Kohn, 'Introduction', in H. T. Buckle, *History of Civilization in England* (New York: Frederick Ungar, 1964), pp. v–xvi.
31. H. T. Buckle, *History of Civilization in England*, new edn, 3 vols (1857; London: Longmans, Green, and Co., 1885), vol. 1, p. 6.
32. Buckle, *History of Civilization in England*, vol. 1, p. 39. This final cause was defined as that which 'produces its principle results by exciting the imagination, and by suggesting those innumerable superstitions which are the great obstacles to advancing knowledge'. Whatever this means, it was clear to Buckle that it was outside of an individual's control.
33. Buckle, *History of Civilization in England*. Particularly relevant is chapter 3, 'Examination of the Method Employed by Metaphysicians for Discovering Mental Laws', pp. 152–67.
34. Buckle, *History of Civilization in England*, vol. 1, p. 156.
35. [Anon.], 'History of Civilization in England', *Crayon*, 4 (1857), pp. 336–8, on p. 336.
36. [Anon.], 'History of Civilization in England', p. 337.
37. [Anon.], 'History of Civilization in England', p. 338.
38. Hacking, *The Taming of Chance*, p. 126.

39. T. M. Porter, 'Lawless Society: Social Science and the Reinterpretation of Statistics in Germany, 1850–1880', in L. Krüger, L. J. Daston and M. Heidelberger (eds), *The Probabilistic Revolution*, 2 vols, vol. 1: *Ideas in History* (Cambridge, MA: Massachusetts Institute of Technology Press, 1987), pp. 351–75, on p. 354.
40. Hacking, *The Taming of Chance*, p. 130.
41. N. Reichesberg, *Der berühmte Statistiker Adolf Quételet, sein Leben und sein Wirken* (Bern: Buchdruckerei Stämpfli, 1896).
42. For a summary of early critics, see Lottin, *Quetelet, statisticien et sociologue*, pp. 436–8. Porter also lists the harshest critics in 'Lawless Society', p. 353, n. 4.
43. G. F. Knapp, 'Die neuren Ansichten über Moralstatistik', *Jahrbücher für Nationalökonomie und Statistik*, 16 (1871), pp. 237–50, on p. 239.
44. Interestingly, two of the most cited works on Quetelet and determinism disagree about the nature of the German response to Quetelet. Porter claims that German statisticians still saw Quetelet as a 'great man', and that Buckle was seen as an impoverished version of the Belgian 'master'. Hacking focuses on the German dislike of Quetelet's ideas alone, in particular Knapp's 'cure' for the 'disease' of *Queteletismus*. In Porter, 'Lawless Society', p. 356, and Hacking, *The Taming of Chance*, pp. 131–2.
45. M. MacLean, 'History in a Two-Cultures World: The Case of the German Historians', *Journal of the History of Ideas*, 49 (1998), pp. 473–94, on p. 475.
46. Kohn, 'Introduction', pp. vii–viii.
47. [J. Herschel], 'Quetelet on Probabilities', *Edinburgh Review*, 92 (1850), pp. 1–57.
48. Turner, *The Search for a Methodology*, pp. 99–104.
49. In linking Quetelet to Buckle, Hacking's 'Nineteenth Century Cracks in the Concept of Determinism', p. 471, claims that Quetelet quoted Buckle 'at length'. While Buckle is given an extended quotation in the 1869 version of *Physique sociale*, the passage refers only to crime, and not the former's claims about statistical determinism. While Quetelet was certainly not holding Buckle at arm's length, it was more his enthusiasm for *any* mention of his theory than an endorsement of the larger claims of *History of Civilisation in England*. Porter's *Trust in Numbers*, p. 167, similarly notes that Quetelet quoted Buckle 'at length' and that Buckle's 'statistical determinism did not offend him [Quetelet]'. It is difficult to disprove a negative claim, but as early as 1842, in the English edition of *Sur l'homme*, Quetelet had vigorously defended himself against charges of determinism, writing that the 'distinctions which I have already established with care in my work, ought to have proved ... how far I am from a blind fatalism'. A. Quetelet, 'Preface', in *A Treatise on Man and the Development of his Faculties*, ed. T. Sibert, trans. R. Knox (French, 1835; Edinburgh: W. and R. Chambers), pp. v–x, on p. x.
50. For the textbook, see L. J. Siegel, *Criminology: The Core*, 3rd edn (2002; Belmont, CA: Wadsworth, 2008), p. 8. Desrosières claims that after 1900, Durkheim, 'never spoke of Quetelet; he ignored him'. A. Desrosières, 'Quetelet et la sociologie quantitative', *Actualité et universalité de la pensée scientifique d'Adolphe Quetelet: Actes du colloque organisé à l'occasion du bicentenaire de sa naissance* (Brussels: Académie royale de Belgique, 1997), pp. 179–98, on p. 193.
51. A. Oberschall, 'The Two Empirical Roots of Social Theory and the Probability Revolution', in G. Gigerenzer, L. Krüger and M. S. Morgan (eds), *The Probabilistic Revolution*, 2 vols, vol 2: *Ideas in the Sciences* (Cambridge, MA: Mahusetts Institute of Technology Press, 1987), pp. 103–34.
52. E. Durkheim, *Suicide: A Study in Sociology*, ed. and trans. J. A. Spaulding and G. Simpson (French, 1897; Glencoe, IL: Free Press, 1951), p. 300.

53. Durkheim, *Suicide: A Study in Sociology*, p. 303.
54. Durkheim, *Suicide: A Study in Sociology*, p. 302.
55. A. Quetelet, *Sur l'homme et le développement de ses facultés*, 2 vols (Brussels: Bachelier, 1835), vol. 2, p. 151.
56. A. Quetelet, *Du système social et des lois qui le régissent* (Paris: Guillaumin, 1848), p. 88.
57. Buckle, *History of Civilization in England*, vol. 1, p. 27.
58. Buckle, *History of Civilization in England*, vol. 1, p. 20.
59. [Anon.], *Examination of Buckle's History of Civilization in England by a Country Clergyman* (London: Thomas Hatchard, 1858), p. 8.
60. F. H. Hankins, *Adolphe Quetelet as Statistician* (New York: Columbia University Press, 1908), pp. 1–136, on p. 82 and Lottin, *Quetelet, statisticien et sociologue*, p. 447.
61. Hacking for instance calls it a 'truly definitive summary and synthesis'. Hacking, 'Nineteenth Century Cracks in the Concept of Determinism', p. 469, n. 37.
62. J. Lottin, 'Statistique morale et le déterminisme', *Revue néo-scolastique de philosophie*, 15 (1908), pp. 48–89, on p. 50.
63. Lottin, *Quetelet, statisticien et sociologue*, p. 496. Ellipses in original.
64. J. Lottin, 'Le Libre arbitre et les lois sociologiques d'après Quetelet', *Revue néo-scolastique de philosophie*, 18 (1911) and Lottin, 'Statistique morale et le déterminisme'.
65. F. H. Hankins, 'Review of *Quetelet, statisticien et sociologue*', *Political Science Quarterly*, 27 (1912), pp. 719–22, on p. 720.
66. Hankins, 'Review', p. 722.
67. Lottin, *Quetelet, statisticien et sociologue*, p. 514.
68. Lottin, *Quetelet, statisticien et sociologue*, p. 515.
69. Oberschall, 'The Two Empirical Roots of Social Theory', p. 116.
70. Hankins, 'Review', p. 720.
71. Hankins, 'Review', p. 722.
72. Hankins, *Adolphe Quetelet as Statistician*, p. 61.
73. Hankins, *Adolphe Quetelet as Statistician*, p. 73.
74. Hankins, *Adolphe Quetelet as Statistician*, p. 78.
75. Hankins, *Adolphe Quetelet as Statistician*, p. 80.
76. Hankins, *Adolphe Quetelet as Statistician*, p. 81.
77. Hankins, *Adolphe Quetelet as Statistician*, p. 82.
78. Hankins, *Adolphe Quetelet as Statistician*, p. 105.
79. As a statistical term, correlation was not coined until the work of Galton and Pearson. It has been argued that even in Galton the idea was not fully developed. S. Stigler, *The History of Statistics: The Measurement of Uncertainty before 1900* (Cambridge, MA: Harvard University Press, 1986), p. 297.
80. Hankins, *Adolphe Quetelet as Statistician*, p. 133.
81. A. Quetelet, *Du système social et des lois qui le régissent* (Paris: Guillaumin, 1848), and A. Quetelet, *Lettres à S.A.R. le duc régnant de Saxe-Cobourg et Gotha: Sur la théorie des probabilités, appliquée aux sciences morales et politiques* (Brussels: Hayez, 1846).
82. Quetelet, *Du système sociale*.
83. Quetelet, *Du système sociale*, p. 105.
84. Quetelet, *Du système sociale*, p. 109. Emphasis in the original.
85. Quetelet, *Du système sociale*, p. 104.
86. Quetelet, *Du système sociale*, p. 109.
87. A. Quetelet, 'Sur la statistique morale et les principes qui doivent en former la base', *Nouveaux mémoires de l'Academie royale des sciences et belles-lettres de Bruxelles*, 21

(1848), pp. 3–68, on pp. 4, 7.
88. P. de Decker, 'De l'influence du libre arbitre de l'homme sur les faits sociaux', *Mémoires de L'Académie royale des sciences, des lettres et des beaux-arts de Belgique*, 21 (1848), pp. 69–92, on p. 73. All emphases – and there were many from the lively philosopher – from the original.
89. Decker, 'De l'influence', p. 74.
90. Decker, 'De l'influence', p. 76.
91. Decker, 'De l'influence', p. 77.
92. Decker, 'De l'influence', p. 80.
93. Decker, 'De l'influence', p. 88.
94. C. Piot, 'Notice sur Pierre-Jacques-François De Decker', *Annuaire de L'Académie royale*, 58 (1892), pp. 215–75, on p. 215.
95. De Decker, 'De l'influence', p. 89.
96. T. M. Porter, 'Lawless Society: Social Science and the Reinterpretation of Statistics in Germany, 1850–1880', in L. Krüger, L. J. Daston and M. Heidelberger (eds), *The Probabilistic Revolution*, 2 vols, vol. 1: *Ideas in History* (Cambridge, MA: Massachusetts Institute of Technology Press, 1987), pp. 351–75, on p. 353.
97. I. Hacking, 'Prussian Numbers, 1860–1882', in L. Krüger, L. J. Daston and M. Heidelberger (eds), *The Probabilistic Revolution*, 2 vols, vol. 1: *Ideas in History* (Cambridge, MA: Massachusetts Institute of Technology Press, 1987), pp. 377–94, on p. 387.

Conclusion: The New Argonauts

1. For the lively debate about the philosophical consequences for these sciences and others, see R. Smith, *Free Will and the Human Sciences in Britain, 1870–1910* (London: Pickering & Chatto, 2013).
2. The connections between Quetelet and Maxwell are best covered in T. M. Porter, *The Rise of Statistical Thinking 1820–1900* (Princeton, NJ: Princeton University Press, 1986), pp. 193–207. On the reaction of French Catholics, see M. J. Nye, 'The Moral Freedom of Man and the Determinism of Nature: The Catholic Synthesis of Science and History in the *Revue des questions scientifiques*', *British Journal for the History of Science*, 9 (1976), pp. 274–92. For a more recent look at Maxwell in particular, see M. Stanley, 'The Pointsman: Maxwell's Demon, Victorian Free Will, and the Boundaries of Science', *Journal for the History of Ideas*, 69 (2008), pp. 467–91, on p. 481.
3. P.-S. de Laplace, *Exposition du système du monde*, 3rd edn (1796; Paris: Chez Courcier, 1808), quoted in M. Crosland, 'A Science Empire in Napoleonic France', *History of Science*, 44 (2006), pp. 29–48, on pp. 37–8.
4. P.-S. de Laplace, *Essai philosophique sur les probabilités* (1814; Brussels: Culture et Civilisation, 1967), pp. 107–8.
5. F. Manuel, *The Prophets of Paris* (Cambridge, MA: Harvard University Press, 1962), p. 312.
6. A. Quetelet, *Sciences mathématique et physique chez les belges au XIXe siècle* (Brussels: Buggenhoudt, 1866), p. 1.
7. J. Cawood, 'Terrestrial Magnetism and the Development of International Collaboration in the Early Nineteenth Century', *Annals of Science*, 24 (1977), pp. 551–87, on p. 551.
8. J. Cawood, 'The Magnetic Crusade: Science and Politics in Early Victorian Britain', *Isis*, 70 (1979), pp. 492–518, on p. 493. Though the 'crusade' failed in its initial attempts to find statistical regularity, it has been argued that it 'established a precedent for government funding of scientific research in Great Britain. E. J. Larson, 'Public Science for a Global Empire:

The British Quest for the South Magnetic Pole', *Isis*, 102 (2011), pp. 34–59, on p. 59.
9. Cawood, 'Terrestrial Magnetism', p. 586.
10. R. K. Pachuri and A. Reisenger (eds), *Climate Change 2007: Synthesis Report* (Geneva: International Panel on Climate Control, 2007), p. 121. For the enthusiasm for these projects, see C. Carter, *Magnetic Fever: Imperialism and Empiricism in the Nineteenth Century* (Philadelphia, PA: American Philosophical Society, 2009). On Herschel's role in particular, see T. L. Hankins, 'A Large and Graceful Sinuosity', *Isis*, 97 (2006), pp. 605–33.
11. G. R. Demarée and T. Rutishauser, 'From "Periodical Observations" to "Anthrochronology" and "Phenology" – the Scientific Debate between Adolphe Quetelet and Charles Morren on the Origin of the Word "Phenology"', *International Journal of Biometeorology*, 55 (2011), pp. 753–61, on p. 756.
12. F. Locher, 'The Observatory, the Land-Based Ship and the Crusades: Earth Sciences in European Context, 1830–50', *British Journal for the History of Science*, 40 (2007), pp. 491–504, on p. 504.
13. See above, Chapter Three.
14. On Quetelet's work in these areas, see É. Brian, 'Transactions statistiques au XIXe Siècle', *Actes de la recherche en sciences sociales*, 145 (2002), pp. 34–46 and N. Randeraad, 'The International Statistical Congress (1853–1876): Knowledge Transfers and their Limits', *European History Quarterly*, 41 (2011), pp. 50–65 and J. L. Heilbron, N. Guilhot and L. Jeanpierre, 'Toward a Transnational History of the Social Sciences', *Journal of the History of the Behavioral Sciences*, 44 (2008), pp. 146–60. Also of value in situating Quetelet in this international network is chapter 3 of J.-G. Prévost and J.-P. Beaud, *Statistics, Public Debate and the State, 1800–1945: A Social, Political and Intellectual History of Numbers* (London: Pickering & Chatto, 2012), pp. 49–62.
15. Charles V, another famous son of Ghent, had tried to use the Low Countries to offset the continental powers. More recent efforts were the work of Otlet, Henri Lafontaine and George Sarton. For a recent study of Otlet that situates him firmly within the Belgian context, see A. Wright, *Cataloging the World: Paul Otlet and the Birth of the Information Age* (Oxford: Oxford University Press, 2014). On Sarton, see L. Pyenson and C. Verbruggen, 'Ego and the International: The Modernist Circle of George Sarton', *Isis*, 100 (2009), pp. 60–78.

Epilogue: The Average Enlightenment

1. M. Hulliung, 'Facing up to the "Enlightenment Wars"', conference paper delivered 14 June 2007, Lyon, France. I owe the phrase 'Enlightenment Wars' to this work. For the published version, see M. Hulliung, 'Rousseau et les Philosophes: Facing up to the "Enlightenment Wars"', in M. O'Dea (ed.), *Rousseau et les philosophes* (Oxford: Voltaire Foundation, University of Oxford, 2010), pp. 235–50.
2. Of the major critics that have posited Enlightenment as a means of control and oppression, see M. Horkheimer and T. W. Adorno, *Dialectic of Enlightenment: Philosophical Fragments*, trans. E. Jephcott (German, 1944; Stanford, CA: Stanford University Press, 2002), J. Gray, *Enlightenment's Wake: Politics and Culture at the Close of the Modern Age* (London and New York: Routledge, 1995) and J. L. Talmon, *The Origins of Totalitarian Democracy* (New York: Praeger, 1961). The critique of Enlightenment thought is of course central to many if not all authors lumped under the problematic heading of 'postmodernism'. Notable critiques I have been induced to read include J. Derrida, *The Archaeology of the Frivolous: Reading Condillac*, trans. J. P. Leavey, Jr (French, 1973;

Pittsburgh, PA: Duquesne University Press, 1980), M. Foucault, 'What Is Enlightenment?', in P. Rabinow (ed.) *The Foucault Reader* (New York: Pantheon, 1984), pp. 32–50 and S. Žižek, *Did Somebody Say Totalitarianism?* (London: Verso, 2001).

3. For deflationary accounts of the Enlightenment project, see R. Darnton, *The Literary Underground of the Old Regime* (Cambridge, MA: Harvard University Press, 1982), D. Outram, *The Enlightenment* (Cambridge: Cambridge University Press, 1995). Outram's 141-page survey of the Enlightenment seems particularly designed to minimize the importance of the subject.

4. For some, it is clearly for worse. Robert Wokler for example has argued that the social sciences usurped the political goals of the Enlightenment, claiming that 'nothing was to prove so destructive of that central feature of the Enlightenment project ... than the birth ... of genuinely modern social science'. R, Wokler, 'The Enlightenment and the French Revolutionary Birth Pangs of Modernity', in J. Heilbron, B. Wittrock and L. Magnusson (eds), *The Rise of the Social Sciences and the Formation of Modernity: Conceptual Change in Context, 1750–1850* (Dodrecht: Kluwer Academic Publishers, 1998), pp. 35–76, on p. 61–2. This extends the central insight of J. Heilbron, *The Rise of Social Theory* (Cambridge: Polity Press, 1995) on the shift in the 'intellectual hierarchy'. For a bizarre vision of the 'better' in the popular discourse, one which traces 'average' people throughout life based on quantitative studies, see D. Brooks, *The Social Animal: The Hidden Sources of Love, Character, and Achievement* (New York: Random House, 2011).

5. Foucault, in fact, seems never to have mentioned Quetelet in all his studies of the sciences of man. Hannah Arendt and others have seen similar evil in monotony, but one would find as many bureaucrats who functioned as enemies of totalitarianism as those who helped make it possible. H. Arendt, *Eichmann in Jerusalem* (New York: Viking Press, 1963) and H. Arendt, *The Origins of Totalitarianism* (New York: Harcourt Brace, 1973). One might just as well view these average men with bemusement as much as horror, as in T. Veblen, 'The Technicians and the Revolution', in T. Veblen, *The Portable Veblen*, ed. M. Lerner (1921; New York: Viking, 1948), pp. 438–65.

6. A détente between intellectual historians and historians of science may be close, with Lorraine Daston recently calling on historians of science to re-embrace ideas. L. Daston, 'Science Studies and the History of Science', *Critical Inquiry*, 35 (2009), pp. 798–813.

7. An interesting work in this regard is the collection J. Golinski, W. Clark and S. Schaffer (eds), *The Sciences in Enlightened Europe* (Chicago, IL: University of Chicago Press, 1999). However, many of the contributions to this work explain *away*, rather than explain, the legacy of Enlightenment.

INDEX

Académie française, 8, 28, 88, 92–3, 98
Académie royale des science, des lettres et des beaux-arts de Belgique, 4, 16, 18, 37, 49, 61–2, 66, 70, 73, 78, 80, 84–5, 87, 90, 95, 102, 112, 116, 129, 156, 160–2
Altona, 78, 102, 108
American Philosophical Society, 7
American Sociological Association, 151
Ampère, André-Marie, 83
anticlericalism, 66
Arago, Dominique-François-Jean, 7, 12, 16, 23, 88–96, 102, 105, 109, 161, 163
art and science (debate), 15, 20–2, 24–7, 31–2, 34–6, 40, 42–3, 54, 59, 62, 65, 86, 106–7
astrology, 76–7
astronomy, 7, 16–17, 43, 55, 59, 68–9, 71–3, 75–6, 86, 89–90, 92, 94, 99–105, 107, 109, 111, 114–15, 123–5, 128, 131, 161–3
  See also observatories (general)
  public, 67, 75, 77–8
Athénée de Bruxelles, 15, 50, 55, 59, 66, 75, 78, 85

Babbage, Charles, 7, 95
Bache, Alexander Dallas, 7
Baily, Francis, 98
Baker, Keith Michael, 27
Becker, Barbara J., 43
Belgian Revolution (1830), 16, 73, 75, 82, 112, 126–7
Belgium, 1, 4–5, 7, 16–20, 24, 35–8, 40–1, 44, 56, 63, 65–71, 73, 75, 78, 80–1, 85–7, 89, 91, 94–6, 98, 101–2, 104–5, 108–9, 114–15, 122, 125–6, 136, 142, 155–6, 158, 161–5

See also Low Countries; *See also* United Kingdom of the Netherlands
education, 15
history, 1, 8, 84, 156, 163
industry, 1, 86
  canals, 82–3, 86
  mines, 82–3, 86
  railroads, 83, 86
intellectual life, 1, 5, 15, 23, 73, 78, 96, 141
science, 1, 8, 17, 43
Belpaire, Antoine, 81
Berlin, 104
Berlin Academy, 84
Berlin Urania, 78
Berral, J. A., 23, 91
biology, 6, 152–3, 159
body mass index *See* Quetelet, Adolphe: body mass index
Bonald, Louis Gabriel Ambroise, Vicomte de, 21–2, 24, 40, 43, 59, 91
Bonaparte, Napoleon, 19, 21, 29, 56, 66, 91, 126
Bourdieu, Pierre, 11–12, 14, 134
Bouvard, Alexis, 7, 12, 16, 73, 75, 81, 83, 88–91, 93–6, 102–3, 105, 109, 112, 126, 161
Brabant Revolution (1789–90), 61
Brown, Robert, 108
Brown, Samuel, 83
Buckle, Henry Thomas, 136, 145–50, 154
  History of Civilization in England (1857), 144–7
Buck, Peter, 5
Buffon, Georges Louis Leclerc, Comte de, 22, 92

Byron, George Gordon, Lord, 45, 68

Cabanis, Pierre Jean George, 26, 28, 34
Cape of Good Hope, 164
Cardano, Gerolamo, 77
Cassini, Jean Dominique (Cassini IV), 72
Cassirer, Ernst, 26
Catholic Church, 21, 29, 39–41, 43, 65–6, 68–9, 82–3, 122, 126, 155–6, 158
Cauchy, François, 81–2, 85–6
Cawood, John, 164
Charles V, 45, 59, 164
Chateaubriand, François-René de, 15, 21, 29–35, 40–1, 43–4, 48–9, 54–5, 59, 68, 92, 158, 161
  influence on Quetelet, 30
Châteauneuf, Louis-François Benoiston de, 121
Christian VII, King of Denmark, 103
climatology, 4, 164
Collins, Randall, 12–14
Commission central de statistique, 81
Comte, Auguste, 6, 22, 28, 32, 38, 41, 135, 145, 147, 153
Condillac, Étienne Bonnot de, 31, 36
Condorcet, Marie Jean Antoine de Caritat, Marquis de, 2, 6, 15, 17, 24, 26–9, 31–2, 60, 65, 71, 88, 90, 97, 113, 126, 145, 161, 163, 167
  influence on Quetelet, 27–8
  social science, 26–8, 124, 159
Copernicus, Nicolaus, 77, 92
Cousin, Victor, 37, 40–2
  eclecticism, 15, 20, 35–42
  sens commun, 40
Crahay, Jacques, 81–2, 86
criminology, 2, 111, 133
Cuvier, George, 88

d'Alembert, Jean le Rond, 24, 26–9, 44, 54, 59–60, 92, 124, 126, 161
Dandelin, Germinal Pierre, 44–5, 53, 55–7, 59, 61–3, 66, 80–2
Darwin, Charles, 135
Daston, Lorraine, 34, 115
David, Jacques-Louis, 44, 60–1
Decker, Pierre de, 17, 136, 154–9, 165, 212
demography, 55, 163

Desrosières, Alain, 5
determinism, 15, 34, 52, 120, 131, 134–6, 143–4, 146–8, 150, 154, 156, 158
Dewez, Louis, Secrétaire Perpétuel, Académie royale de Belgique (1820–4), 80, 84–5
d'Holbach, Paul-Henri Thiry, Baron, 31
Dickens, Charles, 5
Dorpat Observatory, 72
Durkheim, Émile, 9, 135–6, 148–51, 153, 163
  Suicide (1897), 149

earth sciences *See* climatology
eclecticism *See* Cousin, Victor: eclecticism
École polytechnique (France), 79, 88, 91–3, 162
Encyclopaedia (French), 21, 26, 28, 30
Enlightenment, 2, 6, 17, 19–23, 26–8, 30–2, 34, 36–7, 39–41, 46, 52, 59–60, 65, 76, 106, 111, 114, 124–5, 134–5, 157, 161–2, 167–8
  egalitarianism, 2, 167
  philosophes, 6, 22, 26, 28–30, 32, 37, 41, 60, 124, 162, 167
  progress, 1–2, 6, 23, 32–4, 40–1, 49, 51, 59, 67, 94, 101, 111, 116, 162
Esquirol, Jean-Étienne Dominique, 7, 141
evolution, 152, 158

Falck, Antoine Reinhard, 70–3, 79–81, 85–6, 88–90, 105, 131, 161
Faraday, Michael, 7
fatalism, 143, 154
Fonfrède, Henri, 155, 157
Fontanes, Louis-Marcelin, marquis de, 33–4
Fourier, Jean Baptiste Joseph, 7, 27, 88, 109, 116
France, 4, 19–20, 23, 29–30, 32–5, 37–8, 43, 57, 60, 65, 68, 79, 82, 85, 87–8, 90–4, 96, 104, 107, 113, 119, 127, 133, 139, 141, 148, 157, 161–2
  bureaucracy, 21, 29, 34, 62, 65–6, 82, 88, 109, 162
  intellectual life, 21
  science, 34, 88
Frederick VI, King of Prussia, 103

free will, 17, 36, 121, 134, 136, 142–4, 147–8, 150–1, 153–7, 159, 162
French Revolution (1789), 20–1, 57
Fuchs, Stephan, 13–14

Galilei, Galileo, 14, 77, 92
Galton, Francis, 6, 135, 152–3, 163
Garfield, James, 7
Garnier, J. G., 90, 94–6, 99
Gauss, Carl Friedrich, 7, 12, 16, 53, 71, 88, 102–3, 107, 109, 114, 163–4
Gaussian distribution, 7, 88–9, 94, 102, 107, 129
Germany, 4, 9, 13, 16, 19–20, 23, 29, 32–5, 57, 72, 77, 79, 82, 87–8, 102, 104–5, 107–9, 113, 147–8, 157, 161–2
Ghent, 12–13, 15, 18–20, 22–3, 26, 29–30, 44–6, 49–50, 53, 55–9, 62–3, 66, 68–9, 75, 79, 81–2, 90, 92, 96, 98, 122, 126, 156, 161–2
Gillispie, Charles Coulton, 29
Goethe, Johann Wolfgang von, 7, 12, 16, 20, 46, 57, 88, 102, 106–9, 162
Goldman, Lawrence, 6
Gombaud, Antoine, Chevalier de Méré, 114
Graunt, John, 113

Hacking, Ian, 147, 157
Halbwachs, Maurice, 9
Hankins, Frank, 8–9, 147–8, 151–3
Hankins, Thomas, 26
Hapsburg Empire, 1, 19, 23
Heilbron, Johan, 20
Helvétius, Claude Adrien, 111, 135
Herschel, John, 43, 72–3, 83, 87, 148, 164
Herschel, William, 7
Hoffman, Johann Gottfried, 104
Holland, 66, 68–70, 98
    history, 71
    science, 68
Horace, 45, 50, 58–9, 63
Huggins, William, 43
Hugo, Victor, 59
Humboldt, Alexander von, 7, 12, 16, 43, 83, 87–8, 102, 105, 109, 164

ideologues, 20, 23, 26, 28
industry *See* Belgium: industry

Institut de France, 60, 84, 88
Intergovernmental Panel on Climate Change, 17, 164
International Maritime Conference (1853), 17, 162

Joseph II, 23

Kerseboom, Willem, 98, 101
Knapp, George Friredrich, 148
Kuhn, Thomas, 7

Lacroix, Sylvestre, 27
Laensberg, Phillipe, 69
Lalande, Jérôme, 88
Lamartine, Alphonse de, 34–5, 55
Laplace, Pierre-Simon, Marquis de, 24, 28–9, 65, 109, 150, 160
    Essai philosophique sur les probabilités (1814), 96
    Laplace's demon, 143
    See also determinism; *See also* free will; *See also* fatalism
Lavoisier, Antoine, 88
Legendre, Adrien-Marie, 88
Lemaire, Jean-François, 98–9
Le Play, Pierre Guillaume Frédéric, 5
l'homme moyen See Quetelet, Adolphe: average man
Lobatto, Reheul, 98
Lombroso, Cesare, 135
Lottin, Joseph, 9, 147–8, 150–3
Louis XVIII, 29, 60
Low Countries, 15, 17, 19, 23, 44, 59–60, 69, 73, 82, 89, 95, 119, 123, 161, 164

Maclean, Michael, 147
Mailly, Edouard, 26, 45
Mainville, Jacques de, 97
Maistre, Joseph-Marie, comte de, 145, 155, 157
Malthus, Thomas Robert, 7, 117, 123, 125
Manuel, Frank, 6, 28, 162
Marx, Karl, 5, 59
materialism, 26, 36, 38, 40, 52, 135, 147, 152, 154
Maxwell, James Clerk, 7, 135, 159
Medelssohn, Felix, 105

meteorology, 5, 7, 17, 47, 51, 99, 101, 105, 109, 111–12, 163
Mill, John Stuart, 145, 147, 150
Molière (ean-Baptiste Poquelin), 58
Montesquieu, Charles Louis de Secondat, Baron de, 120, 159
Morren, Charles, 116, 164
mortality, 96–8, 101, 116–17, 119–23, 125, 127, 130, 149

Napoleonic Wars (1799–1815), 65, 164
Neptune (discovery), 90, 93
Netherlands, 5, 23, 44, 54, 56–7, 59, 66, 68, 70, 73, 81, 116, 119, 122, 156, 162
Newton, Sir Isaac, 8, 31, 45, 50, 92
Nieuport, Charles, Commandeur de, 60, 80
Nourrit, Adolphe, 126

observational sciences, 117, 125, 127
Observatoire de Paris, 73, 88, 91
Observatoire royal de Bruxelles (later, de Belgique), 4, 16, 66–7, 70, 74, 85
  construction, 16, 54, 58, 62, 73, 75, 78, 81, 102, 115, 132, 164
  instruments, 71, 78, 88, 99, 112
observatories (general), 4, 13, 16, 67, 72, 77–8, 87, 89, 99–102, 105, 116, 156–7, 162
  instruments, 68, 72, 89, 94
  telescopes, 70, 72–3, 77
Olbers, Heinrich, 103–4, 107, 109, 161
Ots, Charles, 45, 56
Ovid, 45, 63, 76

Panama Canal (Belgian Delegation), 81–2
Paris, 5, 13, 16, 20, 27, 37, 53–5, 57, 60–3, 67, 71–3, 77–8, 80–1, 87–91, 93–9, 102, 109, 115, 119–20, 126–7, 142, 161, 163
Pascal, Blaise, 23–7, 29, 31, 45, 50, 54, 59, 114, 161
  esprit finesse and *esprit géomètre*, 25
Pearson, Karl, 6, 135, 152–3
periodic phenomena, 112
physics, 2, 4–5, 7, 11, 17, 20, 25, 71, 73, 75, 93, 96, 111–12, 117, 121, 123–8, 133, 135–6, 144, 146–7, 153–6, 159, 162, 168

Pirenne, Henri, 19–20, 87
Poisson, Simeon Denis, 7, 16, 27, 72, 88, 90, 109, 115, 161
Polignac, Jules de, 68
Porter, Theodore, 147, 157
positivism, 28, 34, 38, 40–2
probability, 9, 15, 24–5, 29, 73, 88, 94–7, 104, 111–15, 119, 124–5, 154, 159, 161

quantification, 5, 7, 50, 163–4
Quetelet, Adolphe
  average man, 1–3, 5, 16–17, 98, 126–7, 130, 133–5, 137–8, 141, 145–6, 149–52, 159, 161, 165, 168
  body mass index, 4, 129–30
  Correspondance mathématique et physique (1825–34), 4, 16, 83, 88, 95–101, 115, 118–20, 157, 162
  critical studies of, 9–10
  criticism, 17, 34, 57, 72, 114–15, 125, 138, 144, 146–7, 152–3
  opera, 15, 44–5, 49–50, 63
    Jean Second (1816), 45–6, 56
    M. Dièse ou l'auteur dans l'embarras (1819), 47–9, 54
    Moschur (1816–17), 46–7, 51
  poetry, 15, 18, 21, 30, 35, 42, 44–5, 48, 50–1, 53–5, 57–9, 62, 76–7, 86, 89, 107
    'À mon ami de Reiffenberg', 52
    'Épître à Dandelin', 53
    'La Poète et la Raison', 54
    'La Poète mourant à son lampe', 50, 195
    'L'Illusion', 52–3
  popular astronomy *See* astronomy: public
  Sciences mathématique et physique chez les belges (1866), 1, 4, 17, 81
  scientific networks, 13–14, 44, 87–8, 111, 161
  social physics, 1–2, 16–17, 112, 123, 125, 127, 136, 153–5, 157, 159–61, 165
  Sur l'homme et le développement de ses facultés (1835), 7, 89, 95, 101, 112, 127–8, 132, 134, 136–9, 143–5, 147, 149, 160
Queteletismus See Quetelet, Adolphe: criticism

Raoul, Louis-Vincent, 44, 50, 55, 57–62, 82
Reichesberg, Naúm, 147–8, 150
Reiffenberg, Baron de, 36–41, 44, 55, 57, 61–3, 80–2, 84, 93
Romanticism, 15, 20, 29–30, 32–5, 37, 41, 43–4, 46, 48–50, 53, 57–8, 76, 106–7, 161–2
  French, 29, 41
  German, 20, 23, 30, 57
Royal Greenwich Observatory, 99
Royal Society of London, 84

Sabine, Edward, 83, 95, 164
Saint-Simon, Comte de, 28, 32
Sarton, George, 7
Schumacher, Heinrich, 103, 109
sciences of man, 2, 6, 115, 134, 157, 167
  See also social sciences
Simons, Pierre, 81–2, 85–6
Skinner, Quentin, 11
social sciences, 2, 5, 7, 75, 113, 135–6, *150*, 153, 160, 167–8
  See also sciences of man
Staël, Madame de (Anne Louise Germaine Necker, Baronne de Stael-Holstein), 15, 21, 23, 30, 32–5, 37, 44, 50, 54, 104, 161
  De la littérature considérée dans ses rapports avec les institutions sociales (1800), 32–4
statistical tables
  births, 16, 99, 104, 116–17, 120
  criminal acts and prisons, 133
  deaths, 97, 104, 116–17, 119–20
  height, 104, 106, 127–33, 139, 141, 162
  population, 42, 75, 81, 88, 96, 98–9, 104, 107, 109, 116–17, 119–20, 123, 127, 130–1, 133, 138, 142, 152, 163–4
  weight, 127–31, 133, 141
statistics, 2, 5, 9–10, 15–16, 25, 41, 55, 59, 65, 73, 75, 88, 96–9, 101, 104, 107, 109, 111–15, 119, 121, 124–7, 133, 142, 144, 147, 151–7, 159, 161, 163
  See also quantification
  correlation, 99, 123, 153
  German, 104, 114–15, 124–5, 161

Stigler, Stephen, 114
Süssmilch, Johann, 113
Switzerland, 142

terrestrial magnetism, 5, 17, 107, 163
Théâtre de Gand, 45
Théâtre Royale de Monnaie, 126
Tocqueville, Alexis Charles Henri Maurice Clerel de, 88
Treaty of Westphalia (1648), 66
Turgot, Anne Robert Jacques, 28, 31, 65, 88, 90

United Kingdom of the Netherlands, 5, 16, 19–20, 42–4, 47, 54, 56–7, 59, 65–7, 69–70, 72, 81, 87, 98, 119, 126, 156, 161–2
  administration, 19, 66, 72, 79, 89, 98, 107, 162
  education, 23–4, 38–9, 43, 75, 86, 93–4, 104, 121–2, 156, 164
  history, 150
University of Louvain, 5, 150
  French occupation, 57, 79

Van de Weyer, Sylvain, 36–8, 41
Van Ewyck, Daniel-Jacob, 71–3, 75, 80, 96, 99, 131, 164
van Hulthem, Charles, Secrétaire Perpétuel, Académie royale de Belgique (1815–20), 79–81, 85
van Meenen, Pierre François, 36, 154–5
Venn, John, 148
Verhulst, Pierre-François, 98
Villermé, Louis-René, 7, 88, 95–9, 109, 120, 122, 144
Virgil, 50, 58, 131
Voltaire, 23, 26–7, 29, 39, 55, 66, 124

Wagner, Adolphe, 147
Waterloo, 19, 126
Weimer, 20, 106
Whewell, William, 7, 108–9
William I, King of the Netherlands, 36, 57, 60, 66, 80, 87–8, 90, 161
Wulf, Maurice de, 36